Beyond a Disenchanted Cosmology

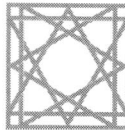

Beyond a Disenchanted Cosmology

Archai: The Journal of Archetypal Cosmology,
Issue 3

Edited by
Keiron Le Grice, Grant Maxwell, and Bill Streett

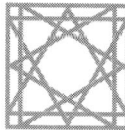

ARCHAI
PRESS

Archai Press
Published by Archai Press
San Francisco, CA
www.archaijournal.org

Series cover design by Kris Wainscott and Darrin Drda

Astrological charts created with Io Edition software from Time Cycles Research (www.timecycles.com)

Printed by CreateSpace in the USA

Acknowledgments

Grateful acknowledgment is made to everyone involved in the production of the *Archai* journal and website from its inception in 2008.

Please address any comments to the editors. You can find our contact details at www.archaijournal.org/contact.

Keiron Le Grice and Rod O'Neal

Archai

The Journal of Archetypal Cosmology

The *Archai* journal explores significant correlations between cyclical alignments of the planets and the archetypal patterns of human experience. Combining a methodology derived from astrology with the archetypal perspective emerging from modern depth psychology, archetypal cosmology is concerned with the analysis of the shifting patterns and cycles of world history, culture, art, and individual biography. Beyond this, archetypal cosmology examines the theoretical basis for these correlations and their implications for the wider world view.

The journal is published annually and features articles by major figures in the field, drawing on scholarship from many other areas such as depth psychology, history, philosophy, cosmology, religious studies, cultural studies, the arts, and the sciences.

The primary aim of *Archai* is to provide a forum for the advancement of archetypal cosmology by disseminating its research and ideas to academia and the wider culture, developing its standards of scholarship, exploring and deepening the archetypal perspective, expanding research into world transit analysis through the study of correlations between planetary alignments and archetypal patterns in world cultural history, and establishing connections between archetypal cosmology and other fields.

Contents

Theoretical Foundations of Archetypal Cosmology

Archetypal Analysis of Cultural History

List of Figures

List of Tables

Introduction

Disenchantment and Integrative Postmodernism

Keiron Le Grice

This issue of *Archai* is concerned, in broad terms, with the rise of the disenchanted world view in the modern era and the possible significance of archetypal cosmology for moving beyond it.

Disenchantment and Desacralization

Disenchantment is a term employed by Max Weber to describe the world view of modern secular society in which, as he puts it, "no mysterious incalculable forces come into play"—a world view in which recourse to mysticism, supernatural powers, gods and goddesses, spirits, and magical explanations is considered both unnecessary and invalid.[1] This disenchanted vision, which gained ascendancy in Western civilization after the Enlightenment and especially during the nineteenth century, was created, Weber observes, by processes of "rationalization" according to which life was to be explained in terms of observable and measurable natural forces. Scientific knowledge increasingly replaced religious belief such that the world, no longer related to a spiritual principle of any kind, was stripped of intrinsic meaning; it was no longer seen as sacred.[2]

Richard Tarnas, in agreement with Charles Taylor, connects disenchantment to the process of *objectification*, by which the external world was viewed not as a meaningful subject, possessing intentionality, interiority, and *telos*, but as an object—as unconscious, lifeless matter moved mechanistically by material forces, that could be measured, controlled, and manipulated. During the modern era, Tarnas points out, the locus of all meaning in the world shifted exclusively to the interior realm of human consciousness. The world itself was conceived as possessing no meaning save for that projected onto it by the human psyche. As he explains:

> Disenchantment, the denial of intrinsic meaning and purpose, essentially objectifies the world and thereby denies subjectivity to the world. Objectification denies to the world a subject's capacity to intend, to signify intelligently, to express its meaning, to embody and communicate humanly relevant purposes and values. . . . This in turn tremendously magnifies and empowers human subjectivity: the felt interior capacity of the human being to be self-defining, self-revising, self-determining.[3]

Mircea Eliade, describing a process he calls *desacralization*, gives a similar assessment, explicitly connecting the rise of the sense of freedom and empowered subjectivity of the modern human with the denial of any transcendent spiritual agency:

> Modern nonreligious man assumes a new existential situation. He regards himself solely as the subject and agent of history, and he refuses all appeal to transcendence. . . . Man makes himself, and he only makes himself completely in proportion as he desacralizes himself with the whole. The sacred is the prime obstacle to his freedom.[4]

Another term—*demythologization*—has similar connotations, referring explicitly to the process by which mythic elements were removed from explanations of the origin, purpose, and functioning of the world.[5] With the rise of science and rational philosophy, myths were no longer seen as true in any sense, but as fictions and falsehoods or as primitive forms of explanation now superseded by scientific knowledge—a view, of course, that still holds sway today. This Enlightenment view of myth, as it is known, is reflected in the *total demythologization* that has occurred over the last two hundred years.

Astrology's Place in Modern Thought

Astrology's fall into disrepute in the modern era is a direct consequence of the rise of the disenchanted world view. In the ancient civilizations of Mesopotamia, Egypt, Greece, and Rome, and through the medieval and early modern periods of Western civilization, astrology existed harmoniously with the cosmology and world picture of the time and, as a result, was generally held in high esteem. In the late modern era, however, with the rise of materialism and mechanistic science, astrology has become marginalized and is no longer central to, or even compatible with, the prevailing conceptions of the nature of reality. Following the discrediting of the geocentric Aristotelian and Ptolemaic cosmology, astrology has failed to enjoy the support of the dominant scientific and philosophical models of the time. Without such support, and with the eventual rejection of all forms of esoteric and supernatural explanations of phenomena after the Enlightenment, astrology has been driven to the periphery of contemporary culture.

Consequently, although astrology remains of historical and cultural interest to scholars, to date it has failed to reestablish itself within academia as a viable cosmological perspective informing the modern world view, having no place in contemporary science, philosophy, or psychology. Deemed a pseudoscience and irrational superstition by some within the academic establishment, the consensus view is that astrology is based on unscientific and erroneous premises, and that its continued existence serves only to demonstrate human gullibility. While this view is partly attributable to the superficial forms of astrology that have become prominent in contemporary culture, it is also, at a deeper level, a consequence of the incongruence between the premises and implications of an astrological world view and the dominant scientific paradigms of our time.

Contrary to the modern disenchanted assumption that we inhabit a world without spirit and without meaning, the astrological perspective points to a meaningful order of creative power and intelligence permeating all things. Directly challenging much of what scientific materialism and secular consumerist society have assumed to be true about the nature of reality, astrology provides an alternative frame of reference from which to make sense of human experience, revealing an archetypal dimension of reality, which the physical sciences for all their extraordinary powers of telescopic and microscopic observation have been unable to disclose. Astrology offers a unique way of interpreting how human experience is shaped by the patterning forces of the universe, by the fundamental energies of nature, and by the archetypal themes that have shaped human life for millennia—and here lies its great value.

A Changing Context: The New Science, Depth Psychology, and Postmodernism

While attempts to accommodate astrology into the dominant paradigm of modern science have been unsuccessful (for reasons addressed in this issue), the emergence of the so-called *new paradigm* perspectives—in physics, biology, psychology, and elsewhere—is slowly bringing forth a new understanding of the nature of the cosmos and of the human psyche that might complement or perhaps even replace the modern world picture. Considered alongside the body of research into archetypal correlations presented in Richard Tarnas's *Cosmos and Psyche* and in the *Archai* journal, this paradigmatic shift makes possible a reevaluation and a reinterpretation of astrology as a core component of a new archetypal cosmology.

The background context to the emergence of the field of archetypal cosmology is the dramatic transformation in the understanding of the nature of reality emerging out of relativity theory and quantum theory in physics during the last century. As is well known, these revolutionary developments cast serious doubts over the most basic assumptions on which classical physics and mechanistic determinism are founded and caused an acute sense of crisis among physicists at that time. Instead of a conception of the universe in which inert material particles existing in empty space are moved mechanistically by external forces according to the laws of Newtonian motion, the universe came to be conceived as a complex, intricately interconnected and interdependent whole. These developments in physics, which have demonstrated the limitations of mechanistic determinism and called into question the traditional concept of scientific objectivity, paved the way for the emergence of the new paradigm scientific perspectives emphasizing holism, organicism, complex causality, field theories, nonlocal connections, and the participatory role of the scientific observer in shaping the phenomena under investigation.[6]

Alongside the revolution in physics, another new theoretical approach, closely aligned with the philosophy of holism and organicism, has emerged out of biology. I refer here to systems theory and to the self-organizational paradigm. Like physics, this perspective has emphasized the significance of patterns of organization and the dynamic process-oriented nature of all kinds of systems, an emphasis it shares with process philosophy, based primarily on the work of Alfred North Whitehead.

As discussed in the previous issues of *Archai*, these important developments in the scientific understanding of the universe have been accompanied by equally significant advances in our knowledge of the human psyche, emerging, most especially, out of the field of depth psychology. In their willingness to seriously engage and seek to understand the meaning of phenomena often dismissed and excluded by conventional approaches—such as spiritual experiences, synchronicities, dreams, fantasies, and the meaningful content of psychopathological conditions—Jungian, archetypal, and transpersonal approaches to depth psychology have challenged the paradigmatic limitations of the dominant modern world view, and disclosed a vaster, more comprehensive picture of the human psyche. Fundamental to this larger vision is the notion of an unconscious dimension to the human psyche, one transcending normal human awareness and giving an a priori order to our conscious experience.

Within the wider cultural context, the development of new paradigm perspectives in science has been accompanied by the emergence of postmodernism—the collective term for those cultural and intellectual movements that critique, challenge, or negate the fundamental assumptions and values of the modern world view. Many truths and assertions that have prevailed in the modern era, and that were once deemed incontrovertible, are now being called into question. Truth, to many postmodern thinkers, is now seen not as absolute fact but as constructed interpretation. Claims to objective validity have been undermined, or at least tempered, by the realization of the inescapable subjectivity of human knowledge. Consequently, postmodernism has permitted a reconsideration of views, philosophies, ideas, and practices deemed incorrect or outmoded in terms of the rational-scientific world view of the modern era. Ancient wisdom traditions, esoteric philosophies, indigenous cultures, and the perspectives of previously repressed minorities are finding a new voice on the postmodern stage. In this climate of change, astrology—in its archetypal form—need no longer be alienated from the wider culture if it can face the challenge of setting forth an explanation of astrological correlations in terms comprehensible to the modern mind.

While the term *postmodern* has become synonymous with deconstruction and the Nietzschean "hermeneutics of suspicion," certain theorists such as Charlene Spretnak, Charles Jencks, and David Ray Griffin have tried to reclaim the term, and to use it to signify not only deconstruction, but also trends towards more integrative and constructive developments that lead beyond the limitations of the dominant modern world view—an emphasis that reflects the original meaning of the term *post-modern*, coined by Black Mountain poet Charles Olson in the mid-twentieth century. Such thinkers have sought to distinguish between late- or ultra-modernism, which they identify with deconstruction, and postmodernism in its constructive sense, which, according to Jencks, is "the continuation of modernity and its transcendence."[7] The modern world view is to be transcended not by the outright rejection of the possibility of there being any world view at all, but rather by overcoming the limitations of modernity by means of revision and movements towards a "synthesizing overview."[8]

One way this transcendence might be achieved is by bringing back that which modernity has discarded as obsolete, to effect a synthesis of the old and the new. Such "double-coding," as Jencks terms it, potentially makes possible a creative fusion of ancient wisdom and modern science by which the inherent limitations and one-sidedness of the modern rational-scientific world view might be overcome and a new integral world view developed.[9] It is just this kind of integral approach that is necessary to understand the astrological perspective.

For archetypal astrology to achieve credibility within any emerging world view, however, it is imperative to meet the challenge of articulating a cosmology (a theory of the nature and structure of the universe) that can lend support to its empirical claims to validity. Closely related to metaphysics, cosmology is a branch of philosophy "that deals with the Universe as a totality of phenomena, attempting to combine metaphysical speculation and scientific evidence within a coherent framework."[10] In a wider sense, of course, a cosmology is far more than just a philosophical conception of reality, for it provides the implicit background to all human actions, the context within which human life takes place. A new cosmology must therefore address the place and role of the human in the universe, in both its physical and psychological dimensions. Both aspects of cosmology are central to the focus of the *Archai* journal and to the field of archetypal cosmology as a whole.

In This Issue

The articles in this issue approach the topic of disenchantment and its transcendence from two overlapping perspectives: conceptual, exploring astrology's relationship to modern philosophical and scientific thought; and as historical narrative, focusing on the psychospiritual, existential, cultural, and cosmological dimensions of Western civilization's journey from the Christian world view through disenchantment to the unfolding spiritual transformation at the turn of the twentieth century and on, finally, to consider the life experience of a prominent modern physicist. These two approaches are addressed in turn within the two main sections of the journal.

Theoretical Foundations of Archetypal Cosmology

The opening essay of this first section is my "Astrology and the Modern Western World View," which considers the main factors, both theoretical and empirical, behind astrology's fall into disrepute in the modern era, before making the case for its re-evaluation in light of the new understanding of the nature of reality advanced in modern physics. This article, taken from my doctoral dissertation, provides the preparatory analysis for the archetypal-mythic world view I outline in *The Archetypal Cosmos* (Floris Books, 2010).

The second paper, "Archetype and Eternal Object" by Grant Maxwell, addresses the privileging, in modern science and philosophy, of the Aristotelian notions of efficient and material causation to the virtual exclusion of formal and final causation. Taking as his starting point David Ray Griffin's attempted synthesis of the ideas of Jung and Whitehead, Maxwell argues that Jung's notion of synchronicity can be interpreted as a rediscovery of formal causation in a modern context. More generally, in agreement with the implications of synchronicity, Maxwell proposes that Jungian archetypal theory, brought into association with Whitehead's process philosophy, can serve as a basis for a new world view restoring interiority and meaning to the world at large.

The final paper of this section, adapted from the appendix of Sean Kelly's *Coming Home: The Birth and Transformation of the Planetary Era* (Lindisfarne Books, 2010), explores the relationship between archetypal astrology and the evolution of consciousness. Kelly identifies the need for an explicit integration of Tarnas's focus on the archetypal dynamics studied in astrology and Hegel's notion of the world spirit acting through history as a *telos* or final cause shaping the unfolding course of planetary evolution. In accordance with his spiral model of the evolution of consciousness, Kelly proposes a synthesis of the archetypal astrological perspective with the uni-temporal goal-oriented trajectory of history suggested by Christian theology, Romanticism, and especially Hegelian philosophy.

Archetypal Analysis of Culture and History

The first article in this section is the final part of Rod O'Neal's detailed archetypal analysis of the Puritan movement. O'Neal focuses here especially on the period of Great Awakening during the 1730s and 1740s, shining a light on spiritual experiences occurring within the horizons of an established Christian world view—but one that was already adapting itself to the rise of science and modern philosophy. O'Neal's article, in addition to its archetypal exploration of the planetary alignments associated with spiritual awakening, addresses the revisions to Puritan theology resulting from the ideas of Isaac Newton and John Locke, which threw down a radical challenge to the traditional understanding of God's involvement in the universe.

The second article, "The Shape of Nihilism" by Joseph Kearns, focuses on the profound cosmological and psychospiritual implications of Friedrich Nietzsche's epochal proclamation of the "death of God" in *The Gay Science* in the early 1880s. Combining a careful, systematic textual analysis with reflections on his own personal experience of nihilism and the shattering of the Christian world picture, Kearns's article takes the reader into the heart of the existential predicament of the modern self, as he considers the host of implications—metaphysical, cosmological, existential, psychological, and moral—resulting from the disenchantment of the modern era. Kearns's essay explores the experience of nihilism as a necessary means to its transcendence, anticipating, through Nietzsche's vision of the *Übermensch*, a post-Christian and post-nihilistic future for the human spirit.

The third article in this section, "The Dark Spirit in Nature," looks at the spiritual transformation that began in the late nineteenth and early twentieth centuries, coinciding with the rare planetary conjunction of Neptune and Pluto. I have commented elsewhere that the "death of God" did not signal the end of spirituality, but its transformation, its rebirth into a different form.[11] In this paper, I consider one of the most important figures in this spiritual transformation: Carl Gustav Jung. In certain respects, Jung's life and work were both a continuation of and a response to themes addressed by Nietzsche. Jung's work was a response, moreover, to a dogmatic Christianity that had grown distant from the spiritual needs of the modern, educated individual, and distant too from the instinctual dynamism of life, which had been increasingly repressed into the unconscious. Coinciding with the fiftieth anniversary of Jung's death in 1961, this article presents an archetypal astrological perspective on Jung's personal relationship to the archetypal dynamics and experiences associated with the Neptune-Pluto complex.

Finally, this issue ends with Clara Lindstrom's insightful and engaging exploration of the life and personality of eminent physicist Richard Feynman. Focusing in particular on the archetypal themes associated with Saturn-Neptune alignment in Feynman's birth chart, Lindstrom shows how this archetypal combination—which is often associated with skeptical atheism, disenchantment, and disbelief—can also manifest as a deep commitment to authenticity forged by a rigorous reality-testing empiricism, a scientifically mediated quest for truth, and a post-Christian secular morality. One implication of Lindstrom's analysis is that archetypal astrology, even when applied within a predominantly secular and scientific context, can illuminate the major patterns and themes of an individual's life, providing a basis for a more deeply meaningful and potentially enchanted relationship with the cosmos.

The order of articles in this section thus reflects the transition from an established Christian world view (Puritanism) through nihilism and disenchantment (Nietzsche) to the rebirth of spirituality in depth psychology (Jung), and then concludes with a study of a modern physicist and the archetypal dynamics underlying his life experiences and philosophy (Feynman). In so doing, this section demonstrates archetypal astrology's capacity to illuminate the underlying dynamics of human experience, whether this experience is interpreted through an orthodox religious world view, a scientific or atheistic perspective, or an esoteric depth psychological model, such as Jung's. Archetypal astrology focuses on the framework of cosmically based meanings within which the evolution of world views and the evolution of consciousness takes place. With its recognition of this background order of archetypal meanings, archetypal astrology might therefore, I believe, significantly contribute to the recovery of a more spiritually meaningful world view in which psyche and cosmos, mind and nature, are seen to reflect a deeper metaphysical order and evolutionary *telos* pervading all things.

Notes

1. Max Weber, "Science as a Vocation," in *From Max Weber: Essays on Sociology,* trans. and ed. H. H. Gerth and C. Wright Mills (1919; New York: Oxford University Press, 1946), 154. For a discussion of this topic and it's relevance to archetypal cosmology, see Rod O'Neal, "Seasons of Agony and Grace: An Archetypal History of New England Puritanism," PhD diss., California Institute of Integral Studies, 2008 .

2. As Weber puts it: "One need no longer have recourse to magical means in order to master or implore the spirits, as did the savage, for whom such mysterious powers existed. Technical means and calculations perform the service. This above all is what intellectualization means" Weber, "Science as a Vocation," 139).

3. Richard Tarnas, *Cosmos and Psyche: Intimations of a New World View* (New York: Viking, 2006), 21.

4. Mircea Eliade, *Sacred and the Profane: The Nature of Religion,* translated by Willard R. Trask (New York: Harvest Books, 1968), 202.

5. For a discussion of demythologization, see Lauri Honko, 'The Problem of Defining Myth,' (1972) in *Sacred Narrative: Readings in the Theory of Myth,* edited by Alan Dundes (Berkeley: University of California Press), 1984. For the relationship of mythology to the modern world view, see Keiron Le Grice, *The Archetypal Cosmos: Rediscovering the Gods in Myth, Science and Astrology* (Edinburgh: Floris Books, 2010), 25–33, 275–280.

6. I address the relevance of these new paradigm perspectives to archetypal cosmology in Part 2 of *The Archetypal Cosmos,* which presents a synthesis of ideas from systems theory and modern physics with Jungian depth psychology.

7. Charles Jencks, *What is Post-modernism?* (Hoboken, NJ: John Wiley & Sons, 1996), 13.

8. See Charles Jencks, ed. *The Postmodern Reader* (London: Academy Editions, 1986), 31.

9. Jencks, *Postmodern Reader*, 12. Jencks himself sees the emergent sciences of complexity and the dialogues between East and West, ancient and modern, and mainstream and marginalized perspectives as central to postmodernism. The ideas under discussion in archetypal cosmology are generally in accord with this larger constructive postmodern movement. The work emerging out of transpersonal theory, the new sciences, deep ecology, participatory research, and inter-religious dialogue, for example, has already made significant contributions to this aim. See also Jencks, *What is Post-modernism?*, 20.

10. Anthony Flew, editorial consultant, *A Dictionary of Philosophy*, 2nd revised edition (London: Pan Books, 1984)78.

11. See Le Grice, *Archetypal Cosmos*, 44–54.

Bibliography

Eliade, Mircea. *The Sacred and the Profane: The Nature of Religion.* Translated by Willard R. Trask. New York: Harvest Books, 1968.

Flew, Anthony, ed. consultant. *A Dictionary of Philosophy.* 2nd Revised Edition. London: Pan Books, 1984.

Honko, Laurie. "The Problem of Defining Myth." 1972. In *Sacred Narrative: Readings in the Theory of Myth*, edited by Alan Dundes. Berkeley: University of California Press, 1984.

Jencks, Charles, ed. *The Postmodern Reader.* London: Academy Editions, 1986.

Jencks, Charles. *What is Post-Modernism?* Hoboken, NJ: John Wiley & Sons, 1996.

Le Grice, Keiron. *The Archetypal Cosmos: Rediscovering the Gods in Myth, Science and Astrology.* Edinburgh: Floris Books, 2010.

O'Neal, Rod. "Seasons of Agony and Grace: An Archetypal History of New England Puritanism." PhD diss., California Institute of Integral Studies, 2008.

Tarnas, Richard *Cosmos and Psyche: Intimations of a New World View.* New York: Viking, 2006.

Weber, Max. "Science as a Vocation." In *From Max Weber: Essays on Sociology,* translated and edited by H. H. Gerth and C. Wright Mills (1919; New York: Oxford University Press, 1946).

Theoretical Foundations of
Archetypal Cosmology

Astrology and the Modern Western World View

Keiron Le Grice

It says much about the status of astrology in the modern Western world that before one can enter into a serious consideration of the subject one must first address the question of why so many people refuse to accept that there might be any truth or value in astrology whatsoever.[1] Such is its discredited standing today that most intelligent people are unwilling to even begin to entertain the possibility that astrology might have any legitimate claims to validity. This paper explores the origins of this extreme skepticism. By examining some of the core philosophical suppositions and scientific paradigms that form the basis of the dominant Western world view, this article will consider the reasons why astrology has fallen into such disrepute. It will then be possible to make the case for a reassessment of astrology in light of the new understanding of the nature of reality that has emerged over the course of the last century.

Conflicting Opinions

It is a uniquely curious, not to say perplexing, fact that although many of the greatest figures in the history of Western thought have been proponents or practitioners of astrology, in the modern era astrology has been roundly castigated and treated with disdain by the scientific and academic establishment. The definitive proclamation of the modern rebuttal of astrology was issued in 1975 in the form of a public statement signed by a group of 186 scientists, including eighteen Nobel laureates, wishing to make their position unequivocal and lay to rest once and for all any lingering suspicions that there might indeed be a correspondence between planetary positions and human experience:

> We, the undersigned—astronomers, astrophysicists, and scientists in other fields—wish to caution the public against the unquestioning acceptance of the predictions and advice given privately and publicly by astrologers. Those who wish to believe in astrology should realize that there is no scientific foundation for its tenets.[2]

The statement continues:

> One would imagine, in this day of widespread enlightenment and education, that it would be unnecessary to debunk beliefs based on magic and superstition. Yet, acceptance of astrology pervades modern society. We are especially disturbed by the continued uncritical dissemination of astrological charts, forecasts, and horoscopes by the media and by otherwise reputable newspapers, magazines, and book publishers. This can only contribute to the growth of irrationalism and obscurantism. We believe that the time has come to challenge directly, and forcefully, the pretentious claims of astrological charlatans.[3]

Yet as these scientists sought to enlighten and protect the general public, they were, whether or not they intended to, just as assuredly challenging and refuting those earlier thinkers who had found much of value in the astrological perspective. The list of eminent scholars, scientists, philosophers, and writers favorably disposed to this ancient cosmological system, or who laid philosophical foundations for its subsequent development, makes impressive reading: Pythagoras, Plato, Aristotle, Hipparchus, Ptolemy, Plotinus, Proclus, Albertus Magnus, Dante, Aquinas, Ficino, Copernicus, Kepler, Brahe, Galileo, Bruno, Goethe, Emerson, Yeats, and Jung, among others.

Of course, in itself the fact that astrology has been highly valued by prominent figures in the history of ideas tells us nothing of its actual validity, but it does suggest that astrology is worthy of more serious consideration than it usually receives and that we should look more closely at the reasons it is now so widely rejected. Certainly, it is hard to reconcile the fact that these two groups should come to adopt such starkly opposing positions as to the validity of astrology. Until recently the weight of the academic establishment and of consensus opinion has, of course, sided with the group of scientists in their critique and repudiation of the astrological perspective. Modern scientific knowledge is widely understood to be superior to all earlier forms of knowledge such that we have been able to recognize and discard those former errors of understanding about the nature of reality that appeared to support astrological correspondences. Thus, astrology's truth claims are deemed to be fallacious, based, it is supposed, on an archaic understanding that modern science has demonstrated to be without foundation. If this is in fact so—if the modern scientific verdict on astrology is indeed accurate—then even those illustrious luminaries cited above, for all their brilliance in other respects, and their undoubted wisdom, were, in their judgment about astrology, under a serious misapprehension.

More probable, it seems to me, is that the modern scientific West, in the pursuit of rational and scientific certainty, has excluded much from its field of concern, and that the very theoretical paradigm shaping and defining the scientific enterprise invalidates astrological correlations a priori. For certain phenomena—certain fundamental aspects of the nature of reality—lie outside the rather narrow focus of the scientific method and obdurately defy explanation in terms of the rational materialism of the modern era. Indeed, when we look more closely, we find that what is common to those above-named supporters of astrology is that they were all operating outside of the theoretical parameters and tacit philosophical constraints of the modern scientific paradigm. Many, such as the ancient Greek philosophers, pre-dated by centuries the coming of modernity; others, with a Romantic sensibility, excelled in fields outside of the province of science; and others still, Emerson and Jung among them, transcended the paradigmatic boundaries of their time. Even those thinkers who were instrumental in the genesis of the modern world view, such as Copernicus and Galileo, were not themselves enmeshed in this perspective as their scientific followers were to be.

One wonders if—by engaging the very faculties and modes of understanding that have been quite deliberately omitted from modern scientific investigation—these individuals were therefore able to perceive the value and truth of astrology, whereas the modern academic and scientist, constrained by narrow paradigmatic assumptions, are not. For a discerning appreciation of the astrological perspective is dependent not only on rational analysis and empirical investigation but also on imagination, feeling, introspection and intuition, on interior depth and self-knowledge, on the recognition of universals, and on being epistemologically open to other, deeper modes of analysis apart from the quantitative and statistical.

One wonders, also, if the scientists' repudiation applies not to astrology per se, but only to astrology as presented and interpreted, incorrectly, through the modern scientific world picture. For in the modern era, in keeping with the rational materialism of our time, astrological correlations have been interpreted in materialistic and mechanistic terms, construed, that is, in terms of material forces emitted by the planets that causally affect human lives through measurable physical influences—an understanding that, as we will see, appears to be wholly inadequate to the depth and complexity of the astrological perspective. For although astrological "influences" have been posited since ancient times and are part of the traditional astrological imagination, these were usually conceived as subtle forces or energies rather than as measurable physical forces.

In the postmodern era, as scientists and philosophers become increasingly aware of the formerly implicit theoretical assumptions behind modern science and ever more acutely conscious of the limitations these assumptions impose on our understanding, we now have the opportunity to reconsider the validity of astrology. Postmodern reflections on science, and the emergence of the new paradigm approaches within science, afford us another perspective from which to reevaluate astrological correlations. Placed in a new theoretical context, we might perhaps see that it is not astrology that is in error but our perception of it, blinkered by our models and paradigms through which we, in the modern West, have come to interpret the nature of the universe.

Science and the Mechanistic World View

To understand the prevailing perception of astrology, and astrology's disparity with the dominant collective world view, we must consider the scientific and philosophical developments that have shaped the modern understanding of the nature of reality. For in the modern West, as is well known, the Scientific Revolution that saw the birth of classical physics and instigated the rapid rise of science and the modern rational world picture created an intellectual climate in which astrology struggled to survive. Astrology's plight was also significantly influenced by the direction of modern philosophical thought, which first established and then reinforced the dominant scientific world view.[4]

Prior to the Scientific Revolution in the sixteenth century, the accepted cosmological model of the universe was geocentric. The Earth was considered to be stationary at the center of the universe and all the planets, together with the Sun and Moon, were thought to orbit the Earth in circles. This model was known as the Ptolemaic cosmology after Claudius Ptolemy, an astrologer-astronomer from Alexandria, Egypt, in the second century CE, whose theory drew extensively on the ideas of Aristotelian science. Ptolemy's geocentric explanation of planetary motion was accepted by the church. Indeed, Christians later came to believe, as a matter of dogmatic faith, that God had placed the Earth at the center of the universe, and that the human being was master of the world, made in the image of God. The Ptolemaic cosmology supports the geocentric perspective assumed by astrology: the astrological chart is calculated to accurately reflect the position of the planetary bodies of the solar system as seen by an observer on Earth. From the vantage point of the individual looking out into the cosmos, all the planets seem to revolve around the Earth, so as long as the geocentric model of the universe was retained in astronomy the astrological perspective appeared to have objective validity.

However, in order to explain observed irregularities in the brightness, direction of movement, and velocity of the planets in their orbits, the later medieval and Arabic revisions of Ptolemy's geocentric cosmology became inordinately complex. The complexity of geocentric theory could be overcome but only by abandoning the seemingly incontrovertible belief that the Earth was the center of the universe and adopting instead the heretical notion that all the planets, including the Earth, orbited the Sun. Copernicus realized this and in 1543 took the courageous step of publishing a heliocentric theory that he had originally developed over thirty years earlier. Kepler's work (supported by Galileo's observation of the heavens using a telescope) confirmed the heliocentric hypothesis, and, despite staunch resistance from the church and the academic establishment, the Copernican revolution was born.

For astrology, this development marked a decisive point of divergence from astronomy because the astronomical cosmology and the astrological one were now seemingly different and conflicting. The geocentric perspective now had no basis in science and was objectively untrue. Thereafter, astrology had to struggle for survival in an incongruent intellectual climate on account of its conflict with the science of the time. While there were no immediate consequences, the wider impact of the Copernican revolution, in particular the impetus it gave to the rise of the scientific world view, was eventually to further alienate astrology from serious intellectual thought.

For the Christianized world the implications of the displacement of the Earth from the center of the universe were momentous. The new heliocentric cosmology directly challenged the authority of the church and biblical scripture, ushering in a period in which any established authority became subject to critical examination in the light of the emerging power of human reason. In this intellectual climate, rationality was to take center stage. Reason and logic were weapons against the false claims of any supposed authority and it was believed that reason could by itself provide accurate knowledge of the world. Educated people began to rely more on rationality and less on faith, more on science and less on the unquestioned authority of church doctrine.

The central figure of the emerging rational philosophy was the seventeenth century French philosopher and mathematician Rene Descartes whose foundational works *Discourse on Method* and *Meditations on First Philosophy* laid the foundations for Western philosophy and science for the next three centuries. Assuming a starting position of absolute doubt about the reality of his existence, and seeking a firm foundation upon which to build his philosophy and overcome this doubt, Descartes came eventually to the conclusion that "so long as I continue to think I am something."[5] This insight gave rise to his *cogito ergo sum*: I think, therefore I am.

Descartes situated human identity in the thinking rational mind or soul set against the body and the external world. "I knew I was a substance whose whole essence or nature is solely to think, and which does not require any place, or depend on any material thing, in order to exist."[6] Reasoning thus, Descartes drew a fundamental distinction between consciousness or thinking and the external, physical world. The thinking spiritual-mental substance (*res cogitans*) is, Descartes claimed, a completely different kind of substance from the corporeal world of matter (*res extensa*). The human being consists of a physical body that is influenced by an incorporeal mind. According to this view, the material world, initially set in motion by God, is machine-like in its operation and can be objectively described and quantified in terms of mathematical laws. In Cartesian philosophy, although mind is held to be causally interactive with matter it is an absolutely distinct substance. Descartes thus advocates a form of what is known as interactive substance dualism.[7]

With this distinction Descartes helped to establish as credible the belief that the material world could be explained without reference to mind because it is possible, he argued, to remove mind and with it human subjectivity from explanations of the functioning of world. Cartesian dualism thus gave further impetus to the belief in the objectivity of scientific explanation. In distinguishing between the subject of experience (the thinking ego) and the object (the material world "outside" of mind), Descartes helped shape the future of the Western world by establishing the philosophical ground for the subsequent development of modern science.

No less significant for the rise of the modern scientific world view were several earlier philosophical developments prior to Descartes that had set the Western intellectual tradition on a course towards empiricism, concrete realism, skepticism, and nominalism. With respect to these developments, William of Ockham, writing in the fourteenth century, argued that in the pursuit of knowledge the focus of attention should be restricted to concrete individual entities; he rejected the ontological reality of universal principles, arguing that universals exist only as concepts in the mind; and he also postulated that knowledge can only be attained through sense perception, not by reason alone.[8]

Equally important was the contribution of Galileo. In addition to his crucial telescopic discoveries, Galileo proposed that science should focus exclusively on the supposedly objective, measurable qualities of phenomena such as mass, number, and size, ignoring the supposedly subjective attributes such as color or smell. Here, almost two centuries before Kant's critical turn in philosophy, was a fundamental distinction that in time effectively stripped sensible, humanly perceived qualities from the external world. With this "bifurcation of nature," as Alfred North Whitehead describes it, the world of so-called *secondary qualities*, dependent on fallible human perception and interpretation, and the quantifiable reality of the external world of *primary qualities*, which were seen as independent of human perception, were thrust apart.[9] Along with the Cartesian *cogito*, this was a decisive development that eventually banished qualitative considerations from the realm of science. That Galileo was himself a practicing astrologer indicates, however, that a mechanistic understanding of the motions of the planets in astronomy need not preclude the belief in an astrological meaning to the planetary positions, motions, and relationships.

Today, of course, it is the focus on the quantitative measurement of celestial mechanics, established by the separation of primary and secondary qualities, that utterly dominates our understanding of the planetary bodies and their motions. Astrology's emphasis on the possible qualitative, psychological, meaningful significance of the planetary dynamics of the solar system is generally greeted with contemptuous disregard by scientists. Nothing, it is supposed, could be more removed from scientific investigation, nothing so blatantly false.

Meanwhile, as a result of Copernicus's heliocentric cosmology, science faced a new urgent challenge: it needed to explain why falling bodies drop to Earth rather than to some other place. Previously, it had been thought that this occurs because the Earth, owing to its supposedly fixed position at the center of the universe, is the natural place to which objects must fall. Obviously, the advent of the heliocentric cosmology had rendered this view obsolete, and the stage was set for a new theory, and the entrance into the scientific arena of Isaac Newton.

Drawing together Kepler's mathematical laws of planetary motion, Galileo's ideas concerning terrestrial mechanical motion governed by forces, and Descartes atomistic-mechanistic philosophy, Newton formulated a robust mathematical framework for modern science. Newton's theories, published in his *Philosophiae Naturalis Principia Mathematica* in 1687, were to form the basis of classical physics until the twentieth century. His response to the challenge of the Copernican revolution was to formulate the theory of gravity, which, together with the three laws of motion, allowed him to comprehensively explain the workings of the physical universe.

According to Newton's model, the universe consists of solid, indestructible, material particles existing in empty space. These particles were considered to be the basic building blocks of matter: all physical objects were collections of these fundamental particles. Time was believed to exist independently of the physical universe, flowing inexorably on, creating our experience of past, present, and future. Both space and time were unconditionally accepted as a priori conditions of the material world. After Newton, all events came to be understood as the effects of forces acting upon material bodies.

Newton's model is often referred to as *mechanistic* because, following Descartes, the universe is conceptualized as a giant cosmic machine in which all events are triggered as part of a causal chain. Like a machine, the movement of one part causes another part to move, which in turn affects a further part, and so on. This theory of the mechanics of the universe is known as *determinism*, in which every effect is the necessary result of an antecedent cause—i.e., the cause *determines* the effect. (The apple falls from the tree to the ground because of gravity, the billiard ball moves because it is struck by another ball—to give two widely cited examples.) After Newton, the mechanistic paradigm was successfully applied to all areas of science on both a macroscopic and microscopic level and the laws of classical physics were proclaimed as fundamental laws of the universe. Newtonian science thus brought with it a sense of triumphant mastery as a new world view was born in which scientists at the time sincerely believed they had the power to explain anything using these fundamental laws. The theoretical mastery of the Scientific Revolution then gave rise to the practical achievements of the Industrial Revolution, as the new knowledge and power furnished by science was exercised to the full.

Although it was not Descartes's intention, since his vision of reality retained a deistic spiritual basis, it is testimony to the efficacy of Cartesian philosophy to articulate a common human experience of the relationship between mind or soul and the material world that in the late modern era many intelligent people envisage themselves as thinking beings existing in a mechanistic, unconscious, and essentially meaningless physical universe. Newtonian physics, finding philosophical and psychological articulation in the work of John Locke, cemented the Cartesian dichotomy between inner and outer, and subject and object, as it appeared to demonstrate mathematically that all events within the material universe could be explained solely through cause and effect mechanics and that there was no need to take human subjective consciousness into account. The human self thereafter appeared radically separate from the functioning of the material world, distinct in essence from the human body, yet mysteriously interacting with it, Descartes believed, through the pineal gland in the brain. The human self existed in an impersonal, mechanistic world that functioned without any kind of involvement from God, without any inherent purpose or meaning. The Cartesian-Newtonian paradigm, as it is now often called, effected a further separation of the human subject from the world and from nature, a further split between the *interior* psyche and the *exterior* cosmos.[10] The subjective meaning of human experience—human aims, purposes, values, feelings, desires, and so on—now appeared radically distinct from the objective world. As the modern world view gained ascendancy, any sense of human participation in a meaningful, ensouled universe was eliminated. The astrological supposition of a relationship between planetary cycles and human experience now seemed increasingly untenable. In the material and mechanistic terms through which astrology was understood, it seemed in fact that there was nothing more remote and unrelated to human experience than the motions of the distant planetary bodies in outer space. Unsurprisingly, then, after the Scientific Revolution, in a climate in which materialism and mechanistic determinism reigned supreme, there was no place for the explanations of astrology based on the old geocentric Aristotelian and Ptolemaic cosmology or the medieval system of correspondences.

Freedom and Determinism

The success of Newtonian mechanics in physics saw the extension of deterministic, mechanistic, and materialistic explanations across all disciplines. This development was accompanied by a gradual decline in alternative forms of explanation, particularly those based on transcendent universal principles, such as the Platonic Ideas. As we have seen, these universals, once thought to be the organizing principles behind the material world, were understood, in nominalist terms, to be merely mental and linguistic categories—to exist in name only, to pertain only to categories of the human mind and not to the external world. With the ascendancy of the modern world view, all phenomena were understood to be explicable solely in terms of natural causes. Those theoretical models and philosophies based on metaphysical factors were deemed unnecessary.

Nowhere is this theoretical orientation more conspicuous than in the modern conception of human nature. Reflecting the application of mechanistic determinism to all areas of investigation, influential theories in both psychology and biology have made it possible to describe and explain human nature such that astrological explanations now seem antiquated and illusory. Scientific explanations, both of the universe and of human existence, seem to have rendered astrology obsolete. In the modern scientific view, it is unnecessary to postulate the existence of extraneous astrological factors, or metaphysical principles or archetypes, when human life can apparently be explained well enough, we areled to believe, in terms of natural causes, whether of our own biological nature or the effects of external influences.

Behaviorism and psychoanalysis, which are the two traditional schools or "forces" in psychology, and DNA research in genetic biology, each put forward deterministic explanations of human life, explanations that have shaped and informed the popular understanding of the origins of personality and the composition of the self. From the psychoanalytic perspective, as is well known, childhood trauma and unconscious drives, repressed desires and sublimated sexual impulses, and the raging battle between instinctual gratification and one's internalized moral code, are the determining factors behind human existence. In the Freudian view, the human personality is unconsciously conditioned by instinctual drives grounded in human physiology; and the emotional complexes of adult life, as more than a century of psychotherapy has shown, often originate from traumatic experiences of early life.[11]

Whereas Freudian psychology brought to light the instinctual and unconscious psychological determinants of human life, behaviorism, first developed by experimental psychologist John Watson in 1913 and later expanded by B. F. Skinner, bypassed the human psyche altogether, focusing instead on the conditioning effects on human behavior of external causes in the environment. Behaviorists see human nature as a product of the environment and all behavior as the result of conditioned responses to external stimuli.[12] Such a conception makes possible a methodological rigor and precision that has delivered important insights into both human and animal learning and motivation. However, many have rightly objected to this narrow characterization of human nature in which, by adopting what has come to be known as the black box approach, consciousness is treated as a largely unnecessary postulate. Human psychology is thus reduced to the analysis of observable, external stimuli and resulting changes in patterns of behavior such that the complexity of human will, of feelings, moods, reflection, and inspiration is almost completely disregarded. With the emergence of behaviorism, then, people were effectively seen as indivisible isolated units lacking interiority; the Newtonian atomistic model of the nature of reality had here fully established itself in the field of psychology. Controversially applying his theory to entire cultures, Skinner thought that human salvation and the cure to the world's problems lies in the mass conditioning of human behavior through the controlled manipulation of environmental stimuli.[13] Human nature, he argued, can and should be totally molded by external conditioning factors to achieve a better society.

Despite their inherent shortcomings, given the explanatory power of both psychoanalysis and behaviorism there is little wonder that they have been extremely influential on the modern mind and its attempts at self-understanding. More generally, deterministic explanations of human life, emphasizing either *nature* (prior determining causes in human biology) or *nurture* (the effects of the environment, particularly during early childhood, on shaping the human personality) have emerged across many disciplines both from within psychology and farther afield. Neuroscience, for example, has focused on the neurochemical processes of the brain in order to explain human consciousness and behavior; Marxist philosophy highlights the influence of social and economic factors on the human condition; and genetics finds causes in the DNA coding within chromosomes.[14] Thus, we arrive at the modern materialistic and deterministic view of the human being as a genetically programmed, biologically driven, and environmentally conditioned physical organism. In such a view, in which the basic makeup of the human personality seems to be comprehensible in terms of concrete, identifiable causes, the astrological perspective, in which human nature is understood in terms of universal principles associated with the planets, seems both superfluous and invalid. Because there is no mechanistic causal explanation of a correlation between planets and human experience, people find it difficult to understand how astrological correlations might work, and therefore generally reject astrology as untrue.

The belief in the freedom of the willing self (an idea which goes back to the ancient Greeks and Christianity, but which has been more recently propounded by existentialist philosophers and is still today broadly reflected in the mainstream collective world view) is another highly influential factor that has contributed to the general suspicion and incredulity with which astrology is viewed. To many people the notion that the planets and zodiacal signs have some kind of power of influence over our lives seems like an affront to our prized sense of free will and self-determination. The popular misconception of astrology that one's fate is unalterably "written in the stars," and that one's free will is impotent compared to the power of one's ineluctable destiny, seems to deprive human beings of the power of self-determination and, in so doing, to mark a return to an oppressive fatalism, to a universe of inescapable predestination. Astrology appears to contradict the idea that we are free to forge our lives and shape our identities through acts of conscious will, to choose and fashion the life we please, and it is therefore perceived as a threat to the sovereign power of the human self. For some people, understandably, this by itself is reason enough to reject astrology out of hand.

The dominant modern world view is thus subject to the influence of two pervasive yet inherently contradictory perspectives: mechanistic determinism, which implies that everything is determined by a mechanistic chain of prior causes, and the belief in the freedom of human will. This contradiction is not without problematic consequences, as Whitehead points out:

> A scientific realism, based on mechanism, is conjoined with an unwavering belief in the world of men and of the higher animals as being composed of self-determining organisms. This radical inconsistency at the basis of modern thought accounts for much that is half-hearted and wavering in our civilisation.[15]

Astrology, as it is commonly understood, finds itself caught between both these positions: Because both the human personality and the events of our lives can be well understood in terms of prior determining factors, and since it is apparent that the future is determined by our own wits and will power pitted against chance and the environment, there seems to be no place for astrological explanations of human life based on the supposed influences of the remote planets. If we shape our own futures, how can astrological factors also be responsible for determining the events and experiences of our lives? If we can explain the nature of our personality and the origin of our life events in terms of prior causes, biological or circumstantial, how can the planetary positions and cycles determine human character or influence our biographical experiences? These questions and concerns, and others like them, have persuaded many that there is little value or truth in astrology.

Debunking Astrology

From this brief exploration of the philosophical and scientific background of the modern Western world, we can see that many of the dominant, though sometimes tacit, assumptions within our world view have contributed to the widespread skepticism towards astrology. In light of the many factors that appear to contradict the basic postulates of astrology, it is hardly surprising that to the modern mind entrenched, whether consciously or unconsciously, in these assumptions, the possibility of there being astrological correlations between the planets and human experience seems fanciful and astrology's truth claims appear to be totally without foundation.

In the modern world the dominant understanding of the nature of reality remains rooted in the mechanistic materialist paradigm that has been so influential within science. Coupled with the idea that a belief in astrology conflicts with the belief of human freedom, it is, above all, because our world view is derived from an understanding of the universe originating in classical physics that, as a society, we dismiss astrology. Many critiques of astrology rest on an unchallenged acceptance of the Cartesian-Newtonian causal deterministic framework.

These critiques might be classified into two main categories. The first category is concerned with the explanation of astrological "influences" and the absence of a plausible account of how the planets can causally affect human lives on Earth. It includes, as well, the related question of how such planetary influences are translated into human life. Proponents of astrology are challenged to explain the purported relationship between the planets and human lives in terms of the known forces and mechanics of classical physics. This category thus relates to the general issue of causal influence.

The second category relates to empirical scientific evidence: Can astrology be validated by demonstrating scientifically that there are actual correlations between astrological factors and the conditions of human life? In addition to the question of statistical evidence for astrology, however, this category can be extended to include the wider issue of the relationship of astrology to science in general, and the question of whether astrology and science might be able to coexist without outright contradiction.

Other questions relating to the specifics of astrological theory and the historical origins and development of astrology also merit consideration. Why are the planets associated with certain qualities, archetypal meanings, and themes and not others? How were the planets' archetypal associations initially discovered? How were qualitative attributions made to the planets? Such questions are obviously central to any comprehensive treatment of astrology. Within the constraints of this essay, however, we will be primarily concerned with the two main categories of critique; we will consider the direct challenge, both explanatory and empirical, posed to astrology by science.

Beyond the Causal Hypothesis

Perhaps the single most important factor in astrology's repudiation and its subsequent exclusion from the consensus Western world view has been the absence of a satisfactory causal explanation of planetary influence. Astrology is deemed untenable because it cannot be explained scientifically, in linear deterministic terms, in that there is no convincing explanation in terms of any known force of how a distant planetary body can influence human existence on Earth. Discussing possible causal explanations of astrological correspondence, astrophysicist Victor Mansfield, himself sympathetic to astrology's truth-claims, states that of the four known forces in nature that might explain planetary influence, the strong and weak nuclear forces can be discounted as they do not act over long distances. Of the other two forces, he continues, electromagnetism can also be ruled out because "movements of free charges easily shield electric forces, and magnetic forces decrease with distance even more rapidly than gravity."

This leaves only gravity itself, which, Mansfield explains, has also been rejected as an explanation of planetary influence by scientists because "the gravitational forces of the doctor and nurse [at birth] are much greater than anything from the planets."[16]

On first inspection, then, the argument against astrology looks watertight. But notice that the repudiation of astrology hinges on the universal applicability of the Cartesian-Newtonian paradigm, the cause and effect model of existence. If this conception of the world is shown to be limited, incomplete, or perhaps even in error, then astrology should not be refuted on the grounds that it cannot be explained in terms of this model. In this scenario, it becomes conceivable that the workings of astrology, although inexplicable in terms of classical physics, can be understood when approached with a different theoretical paradigm. If so, the rejection of astrology might be premature.

Despite significant challenges from idealist philosophers, Romantics, and others, throughout the eighteenth and nineteenth centuries the mechanistic paradigm of classical physics became increasingly dominant and widely accepted. The efficacy of scientific determinism was powerfully demonstrated by the rampant success of industrialization, and there was no reason to suspect that the fundamental laws of nature on which this paradigm was based would later be called into question.

In the early twentieth century, however, following the formulation of Maxwell's electromagnetic field theory in the late nineteenth century, the advent of modern physics cast serious doubts over the most basic of assumptions on which classical physics is founded and caused an acute sense of crisis among physicists at that time. As Fritjof Capra reports:

> The exploration of the atomic and sub-atomic world brought them
> in contact with a strange and unexpected reality. In their struggle
> to grasp this new reality, scientists became painfully aware that
> their basic concepts, their language, and their whole way of think-
> ing were inadequate to describe atomic phenomena.[17]

At the heart of modern physics two new theories, the theory of relativity and quantum theory, have destroyed the absolute truth claims of many of the fundamentals of classical physics.

Einstein's theory of relativity has dramatically transformed our understanding of space and time. Previously, space and time were thought to have absolute, independent existence. Space was construed as a three-dimensional stage of life, as the unchanging background in which physical events occur and in which objects are situated. Time, likewise, was believed to exist separately from space, independently of the material universe; it was thought that there could be a universally applicable measurement of time. According to Einstein's special theory of relativity, however, this commonsense view of space and time is actually inaccurate. Space and time are not absolute but relative; they are not independent of each other, but are inextricably linked and together they form a four-dimensional space-time continuum. The measurement both of space and time, Einstein maintains, is relative to the observer; we cannot know of an objective reality outside of our viewpoint and thus there can be no universal, objective view of the world and no universal measurement of time. The idea of "simple location," to use Alfred North Whitehead's term, the notion that objects actually exist independently in their own definite regions of space and time, had to be abandoned. The universe was now conceived instead as something akin to a mysterious dynamic process.[18]

Inevitably, because it undermined the scientific basis of the Newtonian world picture and called into question the very idea of an objective reality, relativity theory led to a radical reformulation of the entire framework of physics and it also gave rise to Einstein's epochal insight of the equivalency of mass and energy, which he expressed in the famous $E=mc^2$ equation. Out of this insight came an entirely new understanding of the nature of the material world. In this new vision of the world, as Fritjof Capra explains,

> All particles can be transmuted into other particles; they can be created from energy and vanish into energy. In this world, classical concepts like 'elementary particles,' 'material substance' or 'isolated object,' have lost their meaning.[19]

Because of Einstein's insight into mass-energy equivalence, the notion that the universe is comprised of a fundamental substance, that it is made up of irreducible material particles in the form of atoms, had to be abandoned. This view is reinforced by quantum theory. Exploration of the subatomic world in quantum physics has thus far suggested that the basic constituents of matter are not solid atoms, moving around in a mechanical billiard ball type motion, but quarks, leptons, and gauge bosons (according to the standard model of quantum physics)—minute packets of energy that are complexly interconnected. The phenomenon known as quantum entanglement has suggested that the elementary particles at the quantum level appear to possess a degree of interconnectedness that goes far beyond the classical understanding of connections in space and time. Accordingly, the universe might now be conceived, Fritjof Capra suggests, as an unbroken web of relations composed of patterns of interconnections. It is therefore meaningless, we are told, to consider a subatomic particle in isolation for it can only be understood in terms of its interaction with other systems. In Capra's view:

> Quantum theory has demolished the classical concepts of solid objects and of strictly deterministic laws of nature. At the sub-atomic level, the solid material objects of classical physics dissolve into wave-like patterns of probabilities, and these patterns, ultimately, do not represent probabilities of things, but rather probabilities of interconnections.[20]

The implications of modern physics for understanding the nature of reality are obviously profound. According to Capra, we are now faced with a world of "inseparable energy patterns" rather than a world of solid material objects.[21] The idea of a simple causal chain of events linking one separate material body to another has been replaced by the idea that existence is an undivided, interrelated whole. "Quantum theory forces us to see the universe," Capra asserts, "not as a collection of physical objects, but rather as a complicated web of relations between the various parts of a unified whole."[22]

The Newtonian universe comprised of solid indestructible atoms has thus been succeeded by the incomparably more complex picture disclosed by quantum physics and relativity theory. Although modern physics is far from arriving at any settled conception or unified theory of the nature of reality, these new theories have radically deconstructed the old Newtonian world view. Interpreting the significance of these developments, Allan Combs and Mark Holland have gone as far as to suggest that modern physics is in the process of giving birth to "a new mythos, a new topology of reality."[23] "In quantum theory," they continue, "we recover the view of the world as an unbroken fabric in which seemingly separate events do not occur in isolation but, in fact, form pieces interwoven into a single tapestry."[24]

Reformulating Astrological Relationships

With this context in mind, we can now return to the question of planetary influence. Astrologers, it is supposed, are unable to explain how a planet in the solar system can causally affect human life on Earth. In visualizing this problem, we naturally think of two material bodies (the planet and the human being) existing separately in empty space, a vast distance apart. The causal influence we might imagine in its crudest terms as some kind of force passing in a causal chain between the two unconnected physical objects, from the planet to the person. According to the view of reality that emerges from modern physics, however, we must abandon this way of conceptualizing the problem for several reasons.

First, a planet is not an independently existing material object. What appears to us as solid indestructible mass is actually capable of being converted into energy. Mass and energy are interconvertible. A physical mass, like that of a planet, is better understood as a concentration of energy existing within the total field of energy of the entire universe. As David Bohm, discussing the nature of atomic particles and their relationship to the surrounding energy field, explains:

> The field is continuous and indivisible. Particles are then to be regarded as certain kinds of abstraction from the total field, corresponding to regions of very intense field (called singularities). As the distance from the singularity increases, the field gets weaker, until it merges imperceptibly with the fields of other singularities. But nowhere is there a break or division. Thus, the classical idea of the separability of the world into distinct but interacting parts is no longer valid or relevant. Rather, we have to regard the universe as an undivided and unbroken whole. Division into particles, or into particles and fields, is only a crude abstraction and approximation. Thus we come to an order that is radically different from that of Galileo and Newton—the order of undivided wholeness.[25]

Second, in this "order of undivided wholeness" material bodies are not isolated from the space around them. Planets and people are not absolutely separate; they both belong to the one undivided energy field of the universe. Physical objects do not exist in empty space or in a void;

they exist, as Bohm puts it, in "an immense ocean of cosmic energy."[26] What we think of as empty space is actually, he proposed, a continuous, unbroken field of energy.

Third, according to modern field theory, we cannot draw an absolute distinction between the classical concepts of force and matter. Forces between particles of matter are, says Capra, linked to "the properties of other constituents of matter."[27] This means there is no clear demarcation between the particles that make up material objects and the force between interacting particles, because the force is described as, confusingly, itself an interchange of particles. "Both force and matter," Capra notes, "are now seen to have their common origin in the dynamic patterns which we call particles."[28]

To properly understand astrology, then, I believe we must abandon the popular notion that planets causally influence human experience through physical forces and instead base our understanding of astrological correlations on the holistic interconnected picture of the universe disclosed by quantum physics. As I have argued in *The Archetypal Cosmos*, in which I develop an explanation of astrology in terms of Bohm's interpretation of quantum physics, depth psychology, and systems theory, the astrological relationship between planet and person should not be imagined as a force passing between two separately existing entities; rather, it might be better conceived as the result of their mutual participation in the patterned energy field and self-organizing dynamics of the cosmos.

As part of a unified whole and a vast "dynamic web of inseparable energy patterns" the scientific observer and the phenomena being observed are themselves inextricably related.[29] The view of the universe that has now emerged, first out of relativity theory, and later in cosmology, is, first, that all measurements of space and time are relative to the observer, and, second, that human beings are centered in their own perspectives with regard to the cosmos. In astrology, the inescapable subjectivity of human experience is acknowledged by the assumption of a geocentric, person-centered viewpoint—an Earth-based perspective centered on individual human beings. Astrological charts are symbolic maps of human experience based on the actual physical vantage points of individual people, or the location of events. However, the validity of astrology does not depend on the geocentric model being objectively true. Indeed, both Kepler and Galileo were practicing astrologers and saw no contradiction between astrology and their commitment to a heliocentric universe. Moreover, although we know the universe is not actually geocentric, phenomenologically speaking the geocentric perspective remains valid in that, as cosmological evidence suggests, we are always inescapably centered in our viewpoints with regard to the universe. Ancient astronomers, as cosmologist Joel Primack and Nancy Abrams explain, "were wrong astronomically that the Earth is the center of the universe, but they were right psychologically: the universe must be viewed from the inside, from our center, where we really are, and not from some perspective on the periphery or even outside."[30] This assertion reflects the ontological and cosmological centering of the universe on individual human lives. As Primack and Abrams point out, the universe centers on the human being in a number of remarkable ways: human beings exist at the center of universal expansion; we are at the center of what they call the cosmic spheres of time; we are centered in the scale of magnitude of the universe; and we are each individually the center of our own perspective looking out at the cosmos. Thus understood, the astrological perspective, which uses charts centered on specific individuals or locations on Earth, is actually in broad agreement with the modern cosmological conception of an omnicentric universe—a universe of infinite centers. According to Brian Swimme, modern cosmology has "discovered an omnicentric evolutionary universe, a developing reality which from the beginning is centered upon itself at each place of its existence . . . to be in existence is to be at the cosmic center of the complexifying whole."[31] Centered in our own perspective, then, we live and breathe here on Earth in the context, first and foremost, of our own solar system. For although the immense vistas disclosed by recent telescopic exploration of space now make astrology's focus on the solar system seem decidedly provincial, it remains the case, of course, that the cycles of the planets in our solar system define our immediate cosmological vicinity.

Putting the above reflections together, we can see that in the light of modern physics the account of planetary influence as presented above requires radical revision. The deeper level of reality revealed by subatomic physics does not contain separately existing material bodies situated in empty space, causally linked by forces. Instead, as we have seen, mechanistic cause and effect descriptions of reality must be seen within the context of a new vision of the universe as an unbroken patterned energy field. The cosmic machine built by Descartes, Newton, and their successors has been dismantled by modern physicists, and the strictly deterministic model of the universe is now viewed as a theory with limitations, a theory that is useful in certain circumstances only. The linear-causal arguments against astrology that once appeared so compelling are drawn from an understanding of the universe that has been shown to be limited, or even inaccurate, and the grounds for the repudiation of astrology have therefore fallen away. This is not to say, of course, that astrology must necessarily be accepted as true; only that it is not reasonable to reject astrology because it is inexplicable in terms of the mechanistic paradigm.

We should be clear, as we proceed, that although one cannot use relativity theory or quantum physics to adequately explain astrology, the implications of the theories of modern physics call into question critiques of astrology based on linear, deterministic causal models of planetary influence, which generally presuppose the Cartesian-Newtonian model of the universe that informs classical physics. Relativity theory and quantum physics disclose a very different reality to the Cartesian-Newtonian view, and astrology must be seen in the context of this new reality. Even though it cannot be fully explained in terms of the theories of post-Newtonian physics, then, these theories might provide starting points from which one might begin to explore the possible basis of astrological correlations.

In line with recent developments in modern physics, it seems possible that the relationships between the planets and human experience can be understood not in causal mechanistic terms, but rather as a form of non-causal or acausal correlation. Within quantum physics, the investigation of subatomic particles has uncovered forms of relationship that defy causal explanations, as David Bohm notes:

> It is an inference from the quantum theory that events that are separated in space and are without possibility of connection through interaction are correlated, in a way that can be shown to be incapable of a detailed causal explanation.[32]

It might be that the astrological relationship between the planets and human experience is actually *acausal* in that, although the astrological evidence suggests there is some form of relationship, there appears to be no physical force emitted by the planets causally influencing human experiences—no form of linear, efficient causation passing from planets to people. Rather, it seems possible that there is some kind of deeper connection, some kind of underlying pattern inherent in the cosmos, that connects the celestial with the human order, and it is this underlying order that will be our concern here.

Indeed, what we now typically understand by causality, in the modern scientific sense, is but a partial subset of a more comprehensive understanding of causation as originally set forth by Aristotle. Four basic types of cause were originally identified: the *material* cause, which is the substance of which something is composed; the *efficient* cause, which is the external agent that serves to initiate change (as in the case of one billiard ball striking another thereby causing it to move); the *formal* cause, which is the underlying pattern or form that guides the growth of an organism or flow of events; and the *final* cause, which is the *telos*, aim, and purpose of an entity, event, or process. Of these, only the first two are generally recognized in science, and the efficient cause is closest to what we now understand as causation. We should keep in mind, then, what might be categorized as *acausal*, as falling outside the limits of causal explanations, might include the philosophical ideas of both formal and final causation, and both of these are important for understanding astrological correlations. Indeed, as Richard Tarnas points out, although the relationship between the *physical planets* and human experience might be considered acausal, there appears to be a form of complex underlying causation between the *planetary archetypes* and human experience. He explains:

> While the physical planets themselves may bear only a synchronistic connection with a given human experience, that experience is nevertheless being affected or caused—influenced, patterned, impelled, drawn forth—by the relevant planetary archetypes, and in this sense it is quite appropriate to speak, for example, of Saturn (as archetype) 'influencing' one in a specific way, or as 'governing' certain kinds of experience.[33]

Astrology and the Scientific Method

It is not surprising that attempts to explain astrology in causal mechanistic terms have been unsuccessful. The inadequacy of causal explanations of planetary influence meant that, for the most part, belief in astrology was eradicated, particularly among scientists and intellectuals who could see no rational basis for astrology because it seemed impossible that distant planetary bodies could have any influence on human life on Earth. Meanwhile, logical positivism, the philosophical movement based on the ideas of the so-called Vienna Circle in the early twentieth century, had decreed all metaphysical speculation to be meaningless because unverifiable. Instead, the positivists insisted that all inquiry into truth and the nature of reality should be restricted to what can be investigated in accordance with strict scientific method. Consequently, in academic philosophy questions of language and its meaning now take precedence over metaphysical speculation. By the twentieth century, science and rational philosophy attempted to demystify the world by insisting that only those propositions that could be logically deduced from their premises in accordance with the laws of logic, or expressed in the form of a mathematical equation, or empirically proven in accordance with the scientific method could be accepted as true. Astrology, clearly, did not fall into this category.

In response to the challenge of science, and in accord with this positivistic ethos, empirical evidence has been sought to establish statistically significant correlations between certain astrological variables and some of the more overt conditions of human life. Most famously, the Gauquelin studies investigated the correlation between planetary placements in individual birth charts (the planets that were "rising" on the *Ascendant* (the eastern horizon) or "culminating" on the *Midheaven* (the noon point overhead) and various professions). Revising his earlier skepticism towards many aspects of astrology, in the 1990s Gauquelin stated:

> Having collected over 20,000 dates of birth of professional celebrities from various European countries and the United States, I had to draw the unavoidable conclusion that the position of the planets at birth is linked to one's destiny.[34]

However, despite the impressive results from Gauquelin's research and independent corroboration of the validity of the data from figures such as Hans Eysenck, the significance of these data has been downplayed, and, in the main, subsequent research studies of this kind have met with only limited success. Scientific recognition of the validity of astrology is scant. The vast majority of research studies, employing a variety of different design methodologies, have failed to provide enough widely accepted substantial scientific evidence to support astrology. The academic discipline of psychology, therefore, also flatly rejects astrology's truth-claims. Hence, considering the second category of arguments against astrology, relating to empirical scientific evidence, it would seem that there is little to corroborate the assertion that the planetary positions and cycles have any significant correlation with the events and experiences of human life.

Yet what most scientific researchers into astrology have failed to comprehend and appreciate, I believe, is the essential nature of astrological correlations. For astrology, properly understood, is not concerned with prediction of specific events; nor do astrological symbols reveal the specific, concrete details of human life such as one's career or one's material circumstance. Rather, astrology is based on the relationship between planetary positions and the archetypal meaning of human experience; and it is because of the archetypal, multidimensional nature of astrology that astrological correlations elude the orthodox scientific method of investigation. That is, it is because astrology is concerned with underlying thematic meanings, and not with the prediction of specific, concrete events or the description of specific uniform character traits, that scientific tests are unsuitable, and that they are unable either to disprove or to validate astrological truth-claims.

On this point, I must stress also that astrology cannot be tested objectively without taking into consideration both the researcher and the subject under investigation because the discernment of astrological patterns of meaning is dependent, to a considerable extent, on a person's capacity to recognize and comprehend such meaning. The depth, accuracy, and subtlety of the perception of astrological-archetypal themes in human experience are contingent on one's own depth of self-knowledge. The recognition and interpretation of astrological patterns as they are expressed in human life is an art form, which, like every art, demands not only a certain natural aptitude but also years of devoted practice. One must cultivate the ability to think symbolically, to discern the underlying archetypal meaning within a great diversity of outward forms of expression. One's depth of astrological-archetypal insight depends on one having a developed inner life and having acquired a feel for the different astrological principles. One can then relate to these principles not just intellectually, as one might learn fixed interpretations from a textbook by rote, but instead one can get to know what each principle feels like, emotionally and somatically, from having recognized these principles in one's own life experience. For this experience there can be no substitute. A prerequisite for a deep understanding of the workings of astrology, then, is that one must first enter into the astrological perspective, immerse oneself in it. If one is to discern archetypal meaning, one must develop a mode of perception that James Hillman calls the *archetypal eye* by which one can perceive universals and archetypal themes within diverse concrete particulars. While it is obviously unnecessary for researchers who are validating astrology to develop a high level of astrological competence themselves, it is essential that at the very least research studies show awareness of the archetypal nature of astrological correlations. Tests involving randomly selected subjects and control groups drawn from the general population obviously fail to meet this criterion. It is for this reason that, to the best of my knowledge, no scientific research has ever been conducted that could either substantiate or disprove an archetypal interpretation of astrology.[35]

Nevertheless, if astrologers do in fact make definite concrete predictions of future events (as many do) then these deserve to be tested empirically, in accordance with the scientific method. If astrologers posit the existence of relationships between astrological factors and specific careers, or particular interests, or the success or otherwise of relationships, or claim to be able to determine unchanging traits of personality, then these claims too should be empirically validated. Astrology, practiced in this way, should stand and fall by the accuracy of its concrete predictions and the validity of its personality descriptions. But, regarding *archetypal* astrology specifically, how is one to measure and quantify archetypal meaning? Clearly, the empirical validation of archetypal astrology requires a far more sophisticated and nuanced method of inquiry than is currently provided by either quantitative or qualitative research studies in psychology. That said, experts in particular fields (such as cultural history, the history of science, and religious studies) should in principle be able to adjudicate over claims made by astrological researchers into the character of particular periods defined by sets of world transits, although the negative image of astrology might deter scholars from even undertaking evaluations of this kind.

There are many factors that make scientific research into astrology problematic, although not impossible: the challenge of isolating single astrological variables from the multitude of factors used in astrology; the sheer intricate, interconnected complexity of the astrological perspective, which encompasses every dimension of human experience; the tendency for many astrologers, themselves subject to the theoretical bias inherent in the modern world view, to misconstrue the nature of astrological correlations and to make ill-founded and unjustifiable claims of a literal or predictive nature. Most significant, however, as Tarnas emphasizes, is the *multivalent* nature of the planetary archetypes—the fact that the planetary archetypes can manifest in such a great diversity of ways while still remaining consistent with a core archetypal meaning. For it is this inherent multivalence that makes it impossible to predict correspondences between astrological factors and the specific details of human life.

To reiterate, the planetary archetypes, and indeed all astrological symbols, have the same general set of thematic meanings for all people, but they manifest differently in the specific details of every human life. This is the cause of considerable misunderstanding of astrology. The case of so-called "time twins"—two people born at exactly the same time in the same location—has given rise to the question as to why two such individuals often appear to be vastly different from each other perhaps, for example, having completely different interests. If astrology is valid, the argument goes, then astrological twins must be very similar on all counts due to the fact that their birth charts will be identical. But, of course, having the same birth charts indicates only that the charts have the same archetypal pattern, which can manifest in a wide diversity of ways while still remaining consistent with the accepted astrological meanings of the planetary relationships within the chart. How each person expresses this pattern cannot be determined from the information in the astrological birth chart by itself. Aside from interpretive errors, this archetypal multivalence also accounts for why different astrologers give different, yet perhaps equally valid, interpretations of the same charts. Astrological charts refer only to general archetypal meanings and not to the specific, personal factors of life. This point cannot be overemphasized. As Tarnas, discussing the meaning of geometric planetary alignments or *aspects* between two planets, puts it:

> That a given natal aspect can express itself in a virtually limitless variety of ways and yet consistently reflect the underlying nature of the relevant archetypes is of course not only characteristic of all astrological correspondence but essential to it. Astrology is not concretely predictive. It is archetypally predictive.[36]

This means the same astrological factor, while consistently conforming to an underlying archetypal meaning, can manifest in radically different or even diametrically opposite ways. The planetary archetypes are multidimensional and multivalent creative principles, which, although thematically consistent, give rise to a potentially limitless range of forms of concrete expression. A Mars-Saturn planetary aspect, for example, could manifest both as a pattern of defensive aggression and retaliation or as an inability to express anger and to assert oneself; it could be present in the chart of the endurance athlete engaged in a punishing regime of physical training, or in the chart of someone for whom physical activity is impossible because of restrictive circumstance; it refers equally to people who have disciplined themselves to fight and be aggressive, and to people who have been conditioned *not* to fight and to refrain from aggressive behavior. Although, on the surface, the difference between these opposing forms of expression could not be more marked, on closer inspection we can see that they all partake in a common archetypal meaning. Here we have the Mars principle pertaining to self-assertion, aggression, striving and struggle, fighting, physical energy, anger, and the warrior archetype, in combination with the Saturn principle relating to restriction, limitation, concentrated pressure, discipline, structure, repression, fear, and a sense of inferiority. The person who trains to fight—who actually accentuates and improves their ability to fight—is imposing discipline and structure (Saturn) on the physical and aggressive energies (Mars). The person who refuses (Saturn) to fight or show aggression and anger (Mars), although ostensibly pursuing a totally opposite course, is similarly applying Saturnian discipline to their aggressive, assertive, energetic impulses. We can see, therefore, that both possibilities are archetypally consistent—that is, they both conform to the underlying meaning of the Mars-Saturn planetary combination, and often many of these seemingly opposing patterns of behavior are interchangeably or simultaneously present, such is the dynamic complexity of an archetypal pairing.

To recognize archetypal meaning in widely varying modes of expression one often has to examine more deeply the underlying motivations behind patterns of behavior. In the case of the above Mars-Saturn example, one might find that defensive retaliation is a sign of fighting to protect one's unacknowledged weakness; physical training and muscular armoring may serve to bolster or conceal a fragile ego; the hard-edged disciplinarian, similarly, might be motivated primarily by fear. On the other hand, weak passivity may indicate the unconscious repression of anger; and pacifism—ruling out anger or violence because it is perceived as morally wrong or socially unacceptable—might cloak a deeper fear of facing and expressing anger. To uncover the underlying archetypal meanings of our actions we have to discern just what our motivations are behind these actions. Astrology, in this way, encourages a greater depth of insight and understanding of one's nature, which, in time, can give rise to a penetrating self-knowledge—the prerequisite for psychospiritual development.

With a proper recognition of the archetypal multivalence underlying astrology, one can see that it is perfectly possible for two astrologers to give different interpretations of the same planetary configuration that are equally valid in that they both coherently reflect the underlying archetypal meanings of the planets involved. Tarnas himself gives many examples of this multivalence at work in *Cosmos and Psyche*, and this is one of the distinctive theoretical contributions of his work. A statistical analysis of astrologers' interpretations that fails to take into account the archetypal nature of the planetary alignments will be inadequate, noticing only surface differences rather than the underlying themes connecting ostensibly dissimilar behavior patterns. As Tarnas stresses, to properly understand and assess astrological correlations one must cultivate "the imaginative intelligence . . . that is capable of recognizing and discriminating the rich multiplicity of archetypal patterns" in both individual biography and world history.[37]

*   *   *   *

To sum up, then, if mechanistic determinism is not universally applicable, if there are other types of interconnection between phenomena that appear to coexist with linear causality, then it is possible that astrology and scientific determinism are simultaneously valid. Indeed, archetypal correlations seem to coherently coexist with linear causality in a way that calls into question the notion that all things might be explained in terms of prior efficient causes. From an archetypal perspective, astrology does *not* actually contradict deterministic explanations; nor does a belief in astrology impinge on the freedom of the human will. Rather, astrology offers us another perspective that supplements causal-determinism; it provides a larger and deeper frame of reference, a background context of archetypal meaning that helps to illuminate causal factors and scientific explanations, and to inform our acts of will and conscious decisions. Causal determinism and the archetypal astrological perspective are not mutually exclusive or competing theories but complementary—each illuminates and augments the other.

The dominant scientific understanding of human nature represents only one particular way of looking at things. The scientific view is partial and much has been left out of the picture. Like an extremely focused narrow searchlight, science has brilliantly illuminated certain features of reality, but in so doing it has excluded from view vast dimensions of reality lying outside of the illuminated region. The archetypal astrological perspective provides a compensatory wide-lens view, as it were, an holistic perspective that seeks to understand human life in terms of the interior significance of our place within, and relationship to, the whole solar system. And astrology also provides a deeper view—an X-ray photograph, if you like—of the underlying archetypal factors pervading human experience. Thus, whether we wish to point to acts of will, genetic heredity, circumstance, or childhood experiences as prior determining factors behind human experience, subsuming all these factors is the deeper framework of archetypal meanings revealed by the astrological perspective. With respect to the planetary archetypes in astrology, we are not dealing with one-dimensional causal factors that can be predicted to correlate with certain events, or actions, or forms of behavior; we seem to be faced, rather, with creative living powers, autonomous principles rooted deep in the structure of reality itself. These creative archetypal principles are not mechanically and rigidly deterministic but, as living processes, they manifest uniquely in each life experience, they are expressed differently by different people, they vary according to context and circumstance, and also according to the degree of human self-awareness. One can think of the planetary archetypes as being analogous to the Olympian gods of ancient Greece. And just as one would not expect to be able to predict and control, to isolate and dissect, or to measure and quantify the actions of gods, so the archetypal principles in astrology similarly transcend the narrow methodological framework employed in empirical testing. The potential value and validity of the astrological perspective cannot be revealed to the clinical gaze and austere analysis of positivistic science.

Notes

1. This article is a modified extract of a chapter from my doctoral dissertation, "Foundations of an Archetypal Cosmology: A Theoretical Synthesis of Jungian Depth Psychology and the New Paradigm Sciences" (San Francisco, CA: California Institute of Integral Studies, 2009).

2. Bart J. Bok and Lawrence E. Jerome, "Objections to Astrology," in *The Humanist 35*, no. 5 (September/October 1975): 4–6. The primary motivating force behind the release of this statement was the skeptical agenda of Paul Kurtz, the editor of The Humanist, renowned for his ardent critique of paranormal phenomena. Kurtz, along with astronomer Bok and science writer Jerome, sponsored the 1975 statement, which was also submitted to newspapers across North America. For further context on the controversy surrounding this statement and attempts to debunk Gauquelin's research, see Denis Rawlings account (http://www.psicounsel.com/starbaby.html).

3. Bok and Jerome, "Objections to Astrology," 4–6.

4. For the survey of science and philosophy, I draw especially upon the following works: Richard Tarnas, *The Passion of the Western Mind: Understanding the Ideas That Have Shaped Our World View* (1991; Repr. London: Pimlico, 1994); Leslie Stephenson, *Seven Theories of Human Nature* (Oxford: Oxford University Press, 1974); T. L.S. Sprigge, *Theories of Existence* (London: Penguin, 1985); Brian Magee, *The Story of Philosophy* (London: Dorling Kindersely, 2001); Alfred North Whitehead, *Science and the Modern World* (1925; Repr. New York: Free Press, 1970).

5. Rene Descartes, "Discourse on Method," in *Descartes: Selected Philosophical Writings*, translated by John Cottingham, Robert Stoothoff, and Dugald Murdoch (Cambridge: Cambridge University Press, 1988), 36.

6. Descartes, "Discourse on Method," 36.

7. For a discussion of dualism and monism, see Keiron Le Grice, *The Archetypal Cosmos: Rediscovering the Gods in Myth, Science and Astrology* (Edinburgh: Floris Books, 2010), chapters 5 and 7.

8. See Paul Vincent Spade, "William of Ockham," ed. Edward N. Zalta, in *The Stanford Encyclopedia of Philosophy*. 2006. http://plato.stanford.edu/entries/ockham/ (accessed September 23, 2009).

9. See Alfred North Whitehead, *The Concept of Nature: The Tarner Lectures Delivered in Trinity College (*1920. Repr. Cambridge: Cambridge University Press, 1995). Another aspect of this bifurcation, according to Whitehead, is the separation of our

awareness of the experience of the world from the world itself, which purportedly causes that experience. Whitehead's process philosophy and existential phenomenology represent two of the most significant attempts to overcome this separation.

10. While Descartes was not the first to advance either a philosophy of mind-matter dualism or theories of the mechanistic functioning of the external world (the roots of these reach as far back as ancient Greek speculation), his philosophy, together with Newtonian mechanics, was to shape the dominant world conception and inform the scientific enterprise through the modern era. This world conception is often referred to for convenience as the Cartesian-Newtonian mechanistic paradigm, although such a label obviously cloaks a far more complex situation with multiple historical sources contributing to the dominant philosophical assumptions underpinning the scientific enterprise.

11. See Sigmund Freud, *New Introductory Lectures on Psychoanalysis,* sStandard Edition, translated by James Strachey (1930; Repr. New York: Norton & Company, 1965).

12. B. F. Skinner, *Beyond Freedom and Dignity* (1971. Repr. Indianapolis: Hacket Publishing Company, 2002).

13. Skinner, *Beyond Freedom and Dignity.*

14. For example, according to Marx, "It is not the consciousness of men that determines their being, but, on the contrary, their social being determines their consciousness." Karl Marx, *Selected Writings in Sociology and Social Philosophy (*1963. Translated by T. B. Bottomore. Repr. London: Penguin, 1990), 7.

15. Whitehead, *Science and the Modern World,* 76.

16. Victor Mansfield, "An Astrophysicists Sympathetic and Critical View of Astrology," http://www.lightlink.com/vic/astrol.html (1997; accessed January 27, 2009). Mansfield is here addressing the view of well-known astronomer Carl Sagan. For a discussion of Sagan's misunderstanding of astrology, see also Stanislav Grof, "Holotropic Research and Archetypal Astrology," In *The Birth of a New Discipline. Archai: The Journal of Archetypal Cosmology,* Issue 1 (2009), edited by Keiron Le Grice and Rod O'Neal (San Francisco: Archai Press, 2011), 55–56.

17. Fritjof Capra, *The Web of Life: A New Synthesis of Mind and Matter* (New York: Anchor Books, 1997) 5.

18. Whitehead, *Science and the Modern World,* 49.

19. Fritjof Capra, *The Tao of Physics: An Exploration of the Parallels Between Modern Physics and Eastern Mysticism,* 3rd edition (London: Flamingo, 1992), 90.

20. Capra, *Tao of Physics,* 78.

21. Capra, *Tao of Physics,* 92.

22. Capra, *Tao of Physics,* 150.

23. Allan Combs and Mark Holland, *Synchronicity Through the Eyes of Science, Myth, and the Trickster,* 3rd edition (New York: Marlowe and Company, 2001), xxx.

24. Combs and Holland, *Synchronicity,* xxxi.

25. David Bohm, *Wholeness and the Implicate Order* (Repr. London: Routledge, 2002).124.

26. Bohm, *Wholeness and the Implicate Order,* 124.

27. Capra, *Tao of Physics*, 92.

28. Capra, *Tao of Physics*, 92.

29. Capra, *Tao of Physics*, 92.

30. Joel Primack and Nancy Abrams, *The View From the Center of the Universe: Discovering Our Extraordinary Place in the Cosmos* (New York: Riverhead Books, 2006),133.

31. Brian Swimme, *The Hidden Heart of the Cosmos: Humanity and the New Story* (Maryknoll, NY: Orbis Books, 1996), 85–86.

32. Bohm, *Wholeness and the Implicate Order*, 129.

33. Richard Tarnas, "An Introduction to Archetypal Astrology," www.cosmosand-psyche.com/pdf/IntroductiontoAstrology.pdf (accessed August 4, 2009).

34. Michael Gauquelin, *Neo-astrology: A Copernican Revolution* (London: Arkana, 1995), 24.

35. See, for example, "Astrology and Science, Research results," which provides a summary of ninety-one research studies, of various different types, published in four different journals: *Correlations*, *APP*, *AinO*, and *Kosmos*. A review of the abstracts suggests that none of the studies have shown sufficient appreciation of archetypal multivalance and multidimensionality.

36. Richard Tarnas, *Prometheus the Awakener: An Essay on the Archetypal Meaning of the Planet Uranus* (Woodstock, CT: Spring Publications, 1995), 20.

37. Richard Tarnas, *Cosmos and Psyche: Intimations of a New World View* (New York: Viking, 2006), 70.

Bibliography

"Astrology and Science." http://www.rudolfhsmit.nl/hpage.htm (accessed August 3, 2009).

———.Research results. http://www.rudolfhsmit.nl/d-rese2.htm (accessed August 3, 2009).

Bohm, David. *Wholeness and the Implicate Order*. Repr. London: Routledge, 2002.

Bok, Bart J., and Lawrence E. Jerome. "Objections to Astrology." *The Humanist 35*, no. 5 (September/October 1975): 4–6.

Capra, Fritjof. *The Tao of Physics: An Exploration of the Parallels Between Modern Physics and Eastern Mysticism*. 3rd edition. London: Flamingo, 1992.

————. *The Web of Life: A New Synthesis of Mind and Matter.* New York: Anchor Books, 1997.

Combs, Allan, and Mark Holland. *Synchronicity Through the Eyes of Science, Myth, and the Trickster.* 3rd edition. New York: Marlowe and Company, 2001.

Descartes, Rene. "Discourse on Method." In *Descartes: Selected Philosophical Writings*, translated by John Cottingham, Robert Stoothoff, and Dugald Murdoch, 20–56. Cambridge: Cambridge University Press, 1988.

————. *Meditations on First Philosophy With Selections From the Objections and Replies.* 1986. Edited by John Cottingham. Repr. Cambridge: Cambridge University Press, 2003.

Freud, Sigmund. *New Introductory Lectures on Psychoanalysis.* Standard Edition. 1933. Translated by James Strachey. Repr. New York: Norton & Company, 1965.

Gauquelin, Michael. *Neo-astrology: A Copernican Revolution.* London: Arkana, 1991.

Grof, Stanislav. 2009. "Holotropic Research and Archetypal Astrology." In *The Birth of a New Discipline. Archai: The Journal of Archetypal Cosmology*, issue 1 (2009), edited by Keiron Le Grice and Rod O'Neal (San Francisco: Archai Press, 2011): 50–66.

Le Grice, Keiron. *The Archetypal Cosmos: Rediscovering the Gods in Myth, Science and Astrology.* Edinburgh: Floris Books, 2010.

Magee, Brian. *The Story of Philosophy.* London: Dorling Kindersley, 2001.

Mansfield, Victor. "An Astrophysicist's Sympathetic and Critical View of Astrology." 1997. http://www.lightlink.com/vic/astrol.html (accessed January 27, 2009).

Marx, Karl. *Selected Writings in Sociology and Social Philosophy.* 1963. Translated by T. B. Bottomore. Repr. London: Penguin, 1990.

Primack, Joel, and Nancy Abrams. *The View from the Center of the Universe: Discovering Our Extraordinary Place in the Cosmos.* New York: Riverhead Books, 2006.

Sartre, Jean-Paul. *Being and Nothingness: An Essay on Ontology.* 1943. Translated by Hazel E. Barnes. Repr. Abingdon, Oxon: Routledge, 2005.

Skinner, B. F. *Beyond Freedom and Dignity.* 1971. Repr. Indianapolis: Hacket Publishing Company, 2002.

Spade, Paul Vincent. "William of Ockham," ed. Edward N. Zalta, in *The Stanford Encyclopedia of Philosophy.* 2006. http://plato.stanford.edu/entries/ockham/ (accessed September 23, 2009).

Sprigge, T. L. S. *Theories of Existence.* London: Penguin, 1985.

Stephenson, Leslie. *Seven Theories of Human Nature.* Oxford: Oxford University Press, 1974.

Swimme, Brian. *The Hidden Heart of the Cosmos: Humanity and the New Story.* Maryknoll, NY: Orbis Books, 1996.

Tarnas, Richard. *Cosmos and Psyche: Intimations of a New World View.* New York: Viking, 2006.

———. "An Introduction to Archetypal Astrology." www.cosmosandpsyche.com/pdf/IntroductiontoAstrology.pdf (accessed August 4, 2009).

———. *The Passion of the Western Mind: Understanding the Ideas That Have Shaped Our World View.* London: Pimlico, 1991.

———. *Prometheus the Awakener: An Essay on the Archetypal Meaning of the Planet Uranus.* Woodstock, CT: Spring Publications, 1995.

Whitehead, Alfred North. *The Concept of Nature: The Tarner Lectures Delivered in Trinity College.* 1920. Repr. Cambridge: Cambridge University Press, 1995.

———. *Science and the Modern World.* 1925. Repr. New York: Free Press, 1970.

Archetype and Eternal Object
Jung, Whitehead, and the Return of Formal Causation

Grant Maxwell

At first glance, C. G. Jung and Alfred North Whitehead might seem to have little in common. On the one hand, Jung spent the formative years of his early career working as a psychiatrist in a mental institution, during which time he began his eight-year association with Freud, which resulted in Jung playing a large, and often unacknowledged, role in the early promulgation of psychoanalysis. Over the half-century following his break with Freud, Jung built his own approach to depth psychology that engaged primarily with the psychological reality of fantasy images, myths, and dreams. On the other hand, Whitehead had a full career as an influential mathematician, writing the seminal book on modern mathematics, *Principia Mathematica*, near the beginning of the twentieth century with his student Bertrand Russell. In his sixties, Whitehead made the shift to philosophy, specifically metaphysical cosmology, though he always retained the rigorous and abstract approach characteristic of mathematics, a level of precision that philosophers have often sought to emulate. Thus, on the surface, it would seem that Jung's engagement with psyche and Whitehead's engagement with cosmos have virtually nothing in common. However, as this implicit reference to Richard Tarnas's *Cosmos and Psyche* indicates, the deepest intimations of their psychology and cosmology, respectively, seem to suggest a convergence of these two thinkers' projects. Indeed, as I have read their work over the years, I have increasingly come to see profound connections between their ideas, particularly between Jung's concept of *archetypes*,

Whitehead's concept of *eternal objects*, and what I perceive as their mutual association with formal causation.

David Ray Griffin's Synthesis of Jung and Whitehead

I was already familiar with David Ray Griffin as an editor of Whitehead's *Process and Reality*, and when I came across *Archetypal Process*, a collection of essays edited and introduced by Griffin, the connection between Jung and Whitehead that I had suspected seemed to be confirmed. This book, published in 1989, emerged out of a conference at the Center for Process Studies at Claremont University Center and Graduate School in 1983 at which Griffin, James Hillman, and other notable Jung and Whitehead scholars met and presented papers.[1] Reading Griffin's long introductory essay, I was generally in agreement with his inaugural attempt to integrate the ideas of Jung and Whitehead, though there were a number of points where I disagreed with Griffin's interpretation, particularly in his portrayal of some of Jung's concepts. As Griffin notes:

> This introduction does not pretend to be written from a neutral or transcendent perspective. Although I have been greatly informed and enlarged by my encounter with archetypal psychology, I approach the question of the relation between the two movements with the sensibilities, interests, and biases of an advocate of process theology.[2]

Thus, while Griffin should be commended for initiating the synthesis of Jung's and Whitehead's work, as well as for recognizing the limitations of his own subject position, his preliminary synthesis does seem to require some significant modifications. Nevertheless, any further attempt to integrate the ideas of these two thinkers must take the considerable accomplishments of *Archetypal Process* into account. Therefore, using Griffin's introduction as my basis, I will endeavor to clarify some of the points where Griffin is unclear and, in the process, begin to show Whitehead's relevance to the discipline of archetypal cosmology, which, as the name implies, has tended to rely more explicitly on Jung's thought than on Whitehead's.

Griffin's contrasting of Jung's and Whitehead's approaches is clear and says much of what needs to be said by way of articulating Whitehead's relevance to archetypal cosmology:

> Whitehead's approach is avowedly philosophical, even metaphysical; Jung contrasts his empirical approach with philosophy, and disdains metaphysics. Whitehead sought to return to pre-Kantian modes of thought, circumventing the Kantian critique by correcting some errors in pre-Kantian philosophy; Jung's primary philosophical sources are Kant himself and post-Kantian philosophers, especially Schopenhauer and Nietzsche. Whitehead deals primarily with concepts, Jung with images. Whitehead is concerned primarily with cosmology, only secondarily with the human soul as a completion of the cosmology; Jung is concerned primarily with the soul, only incidentally with cosmology as the context of the soul. Whitehead employs the impersonal criteria of self-consistency and adequacy to the widest possible range of evidence, seeking to overcome personal bias and limitations of experience; Jung bases his thought largely on his own inner experiences.[3]

Indeed, the two thinkers' approaches can partially be summed up by mere reference to their initial chosen professions: mathematician and psychiatrist. Whitehead's approach was primarily through the logical relations of abstract symbolic language while Jung's approach was based on his extensive practical experience exploring the images produced in the psyche. It should be acknowledged at the outset that these are both entirely valid ways of understanding the complexity of experience and, as I will attempt to show, they are highly complementary to the point of being necessary to one another for the sake of completeness.

Relevance to Archetypal Cosmology

Our first concern here is Jung's and Whitehead's mutual relevance to the discipline of archetypal cosmology and the astrological perspective on which it draws. Based on the above passage from Griffin, the correlation could not be more direct. As implied earlier, the "archetypal" part of the very name of the discipline is predominantly Jungian. However, as Griffin aptly notes, Jung was only tangentially interested in the second part of the discipline's name, "cosmology," whereas this subject was one of Whitehead's primary philosophical concerns. As Keiron Le Grice writes in *The Archetypal Cosmos*, "what has become clear is that we cannot use Jungian psychology to formulate an explanation of astrology without considering the wider philosophical framework within which Jung's ideas are situated," a project to which Whitehead's philosophy is particularly well suited.[4] Thus, as Griffin articulates it, "archetypal psychologists could acquire from process theology a philosophical-theological framework that is compatible with scientific evidence and the facts of ordinary experience as well as with the somewhat extraordinary dimensions of experience presupposed and focused on by archetypal psychology."[5] While this is not quite how I would articulate the relationship (as Griffin's rhetorical formulation, both here and elsewhere, seems to perhaps subtly diminish Jung and exalt Whitehead), the implication appears fundamentally sound: Whitehead's systematic metaphysical cosmology can provide a philosophical grounding and justification for Jung's psychology of archetypes. Whitehead's work is one of the best candidates to account for the philosophical issues at the root of Jung's ideas, an area where Jung was admittedly hesitant to go. Similarly, Jung's work can address the psychological, imaginal, and mythic domains of experience, which Whitehead was less inclined to explore.

This complementary relationship is likewise given sanction by Tarnas in *Cosmos and Psyche*, which, like Darwin's *The Origin of Species* for evolutionary biology or Freud's *The Interpretation of Dreams* for psychoanalysis, is the foundational text of archetypal cosmology, providing a body of historical evidence and a philosophical framework for the field. It is significant, therefore, that the two epigraphs at the beginning of the first chapter of Tarnas's book are from Jung and Whitehead, implicitly positioning these thinkers as two of the primary antecedents to the archetypal cosmological perspective:

> In each age of the world distinguished by high activity, there will be found at its culmination, and among the agencies leading to that culmination, some profound cosmological outlook, implicitly accepted, impressing its own type on the current springs of action.

—Alfred North Whitehead, *Adventures of Ideas*

Our psyche is set up in accord with the structure of the universe, and what happens in the macrocosm likewise happens in the infinitesimal and most subjective reaches of the psyche.

       —C. G. Jung, *Memories, Dreams, Reflections*[6]

These epigraphs, like the quote from Virgil that Freud uses as the epigraph for *The Interpretation of Dreams*, set the tone for the entire work and, thus the discipline that it largely initiated. Through the juxtaposition of these two passages, Tarnas implies that our psychological insight, the way we view the cosmos, and our "cosmological outlook," the cosmos that we view, are profoundly imbricated. As Tarnas writes later in the same book, "world views create worlds"[7] just as much as worlds create world views, for not only does the human mind emerge from the cosmos, but the implicit world view on which the mind bases its relation to experience also conditions the type of meaning that can be "projectively elicited" from the world.[8] Still later, Tarnas writes: "archetypes thus can be understood and described . . . in Whiteheadian terms as eternal objects and pure potentialities whose ingression informs the unfolding process of reality," which is one of the primary points that this essay will elucidate.[9]

In addition to these textual sanctions for the synthesis of the Jungian archetypal perspective and the Whiteheadian cosmological perspective, Tarnas co-taught a seminar with cosmologist Brian Swimme as part of the Philosophy, Cosmology, and Consciousness (PCC) program at the California Institute of Integral Studies in 2007 entitled "Archetypal Process: Whitehead, Jung, and the Meeting of Psychology and Cosmology," which directly prefigures the synthetic project of this essay.[10]

Archetypes and Eternal Objects

In his introduction to *Archetypal Process*, Griffin suggests that both Jung's concept of archetypes and Whitehead's concept of eternal objects "reassert something like the Platonic view of the importance of formal causes in the nature of things."[11] I wholeheartedly endorse this view and commend Griffin for making the connection between these two seminal thinkers explicit. However, I would like to dispute Griffin's assertion that synchronicity "is probably the weakest element in Jung's speculations" and argue that, in fact, Griffin's dismissal of synchronicity in favor of what he terms Whiteheadian "panexperientialism" betrays a fundamental misconception by Griffin of what synchronicity constitutes.[12] In all fairness, this confusion is understandable since Jung himself often seemed ambiguous about the ontological status of the archetypes and their relationship to synchronicity. Moreover, the subtitle of Jung's *Synchronicity*, "an acausal connecting principle," is rather misleading since synchronicity is perhaps better conceived, I will suggest, as a modern psychological inflection and renomination of formal causality as a reaction to the privileging of material and efficient causality in modernity. Nevertheless, I will argue that Tarnas's explication of the nature of synchronicity and the archetypes in *The Passion of the Western Mind* and *Cosmos and Psyche*, informed particularly by the work of James Hillman, sheds light on the conceptual error at the root of Griffin's misreading. Furthermore, I will suggest that synchronicity and panexperientialism, far from being competing explanations of the "parapsychological phenomena" that Griffin discusses, are, in fact, complementary concepts, both expressing different aspects of formal causality. Indeed, though Griffin does not specifically discuss astrology, he does show how both Jung's and Whitehead's work can provide justification for other kinds of "nonlocal" "action at a distance" (the property of certain entities in quantum mechanics by which these entities "communicate" information to one another instantaneously without any apparent exchange of energy), which, as Le Grice suggests in his discussion of quantum entanglement in the previous article, are applicable to the archetypal astrological perspective.[13]

To give some background, near the beginning of the philosophical enterprise, Aristotle, according to the *Dictionary of the History of Ideas*, "distinguished four senses of 'cause': the material out of which things come; the form which things eventually have when they are perfected; that which brings about this completion, the moving [or efficient] cause; and finally the purpose or function of such things, the final cause."[14] Although material causality, which implies efficient causality, has been privileged in modernity, both Jung and Whitehead have argued, in different valences, for the reinstatement of formal causality, which implies final causality. This reinstatement of the repressed modes of causality alongside material and efficient causality seems to be the only way to render the multivalent complexity of experience intelligible.[15] As Griffin articulates it, both Jung and Whitehead rejected

> the modern view of causation. Of the Aristotelian four causes, I have already mentioned both men's rejection of modernity's limitation of efficient causation to contiguous events, and its limitation of material causation to the energy embodied in the entities studied by physics (as each affirms a greatly expanded notion of energy or 'creativity'). But modernity also had a very limited notion of formal causation: the forms embodied in things were limited to mathematical forms. And final causation, or teleology, was eliminated altogether.[16]

Griffin continues:

> Jung and Whitehead both reassert something like the Platonic view of the importance of formal causes in the nature of things. Jung does this, of course, by making archetypes central. The technical term for formal causes in Whitehead's thought is 'eternal objects,' and he explicitly affirms that, besides eternal objects of the objective species (the mathematical Platonic forms), there are also eternal objects of the subjective species (PR 291), which include anything that can qualify the subjective form of a feeling, such as emotions. But also the whole panoply of metaphysical principles, which Whitehead calls the 'categoreal scheme' (PR 18-29), must be regarded as eternal formal causes of everything that occurs.[17]

Thus far, I agree with Griffin's assessment that the Jungian archetypes, in their most mature formulation, seem to be intrinsically related to the Whiteheadian eternal objects by their mutual association with formal causation, though approached from the somewhat different perspectives of psychology and metaphysics. As Jung writes in *Synchronicity*, "the archetypes are formal factors responsible for the organization of unconscious psychic processes," which seems to indicate a direct relation between archetypes and the Platonic forms from which formal causation derives its name.[18] Similarly, I am in agreement with Griffin's assertion that

> Finalism or teleology is equally affirmed by both. One of Jung's major divergences from Freud was due to the latter's attempt to explain all human experience in terms of efficient causes, whereas Jung became convinced that our aim toward the future—our aim to individuate, or realize our Self—was equally important. Whitehead equates 'mentality' with final causation, meaning self-determination in terms of a momentary goal, so his ascription of a 'mental pole' to each actual entity is an ascription of final causation to all actual entities. He in fact says that each actual entity embodies a 'subjective aim.'[19]

Leaving aside Griffin's slightly misleading attribution of the pronoun "our" to the Jungian concept of Self, as the Self is the transpersonal formal source and final goal of the psyche's individuation process, he seems to be fundamentally correct that both Jung and Whitehead affirmed teleology. In fact, because formal causality directly implies final causality (the formal potentiality, impelling from the transtemporal origin, implies its teleological end, pulling towards its final satisfaction), it logically follows from their mutual positions on formal causality that both Jung and Whitehead would be in favor of the reinstatement of final causality. That is, the fact that there are archetypal potentialities of being implies that these formal eternal objects are teleologically pushing towards an actualization of those potentialities, though the specific forms in which they manifest are not determined, like the Aristotelian-Hegelian acorn that teleologically calls forth the oak in all its unforeseeable particularity, or the embryo that bodies forth two genetically identical, though developmentally differentiated twins. In light of Whitehead's thought, and particularly of what Griffin refers to as "panexperientialism" (see below), this developmental impetus constitutes what might be described as a lower octave or fractal reiteration of the emergent quality of mind, which seeks what Whitehead calls "final satisfaction" for its "subjective aims."[20]

One point on which I differ from Griffin is in what seems to be his direct, one-to-one correlation of the eternal objects and the archetypes.[21] Rather, the archetypes seem to be a subset of the eternal objects at their most complex level. For instance, Whitehead writes that "qualities . . .

such as colours, sounds, bodily feelings, tastes, smells, together with the perspectives introduced by extensive relationships, are the relational eternal objects whereby the contemporary actual entities are elements in our constitution."[22] While the qualities named by Whitehead are generally associated with particular archetypes, it does not seem to be the case that archetypes as described by Jung and refined by Hillman, Tarnas, and others, correlate on a one-to-one basis with these fundamental qualities of experience. Rather, archetypes seem to correlate more closely with "the perspectives introduced by extensive relationships" that organize the more basic, fundamental eternal objects that Whitehead mentions. As Whitehead defines the general scope of his concept, "any entity whose conceptual recognition does not involve a necessary reference to any definite actual entities of the temporal world is called an 'eternal object.'"[23] Thus, on the one hand, any potentiality, such as an archetype, that is not related to a particular temporal occasion is necessarily an eternal object, for occasions can only change in their particular temporal manifestations, not in their eternal, a priori form. On the other hand, as Hillman explains in *Re-Visioning Psychology*, "the archetypal perspective offers the advantage of organizing into clusters or constellations a host of events from different areas of life," which seems to suggest that archetypes are more complex agglomerations of qualities than the simple qualities enumerated by Whitehead.[24] Later in the same book, Hillman writes:

> As Jung refined his insight into these complex persons, the persons of our complexes, he discovered that their autonomy and intentionality derives from deeper figures of far wider significance. These are the archetypes, the persons to whom we ultimately owe our personality. In speaking of them, he says that "we are obliged to reverse our rationalistic causal sequence, and instead of deriving these figures from our psychic conditions, must derive our psychic conditions from these figures. . . . It is not we who personify them; they have a personal nature from the very beginning." . . . *We are always talking about persons even at the most abstract level of discussion,* for these foundations, too, are archetypal persons."[25]

Archetypes are modes of consciousness, "best comparable with a God," that orient our relation to the world and our perception of it in particular domains of discourse, for "all ways of speaking of archetypes are translations from one metaphor to another."[26] Thus, while metaphor constitutes the meaningful connection of elements on different levels of experience, the basic eternal objects that Whitehead mentions—"colours, sounds, bodily feelings, tastes, smells"—are not intrinsically metaphorical, though they are susceptible to metaphorization, to coin a term, when they are subsumed into emergent archetypal fields of meaning. For instance, in addition to the classification of archetypes as eternal objects by both Tarnas and Griffin, Le Grice notes that the archetypal "Plutonic dimension of experience encompasses . . . those universal qualities or 'eternal objects,' to use Whitehead's term, that are associated with the underworld theme: repugnant smells, foul tastes, the colour black, extremes of heat and cold, and so on,"[27] which seem coextensive with the singular qualities mentioned by Whitehead, as opposed to the more complex "extensive relationships." Furthermore, according to Whitehead, "there is not, however, one entity which is merely the *class* of all eternal objects. For if we conceive any class of eternal objects, there are additional eternal objects which presuppose that class but do not belong to it."[28] Thus, the archetypes seem to be one "class" of eternal objects that are presupposed by, but not reducible to additional, more fundamental eternal objects. Where the eternal objects constitute anything whatsoever that is pure potentiality unmanifest in temporality, the archetypes, for Hillman, are more specifically personified modes of potential meaning, applicable, like metaphor, on many levels of experience. As Hillman asserts, even when the archetypes manifest in abstract elemental processes, they are always already associated with archetypal persons. Thus, Saturn, for instance, is associated with dryness, but also with slowness, distance, old age, conservatism, focus, rigor, and so on. Each of these individual characteristics of the Saturn archetype is an eternal object that, combined, synthesize to form the emergent Saturn archetype, which can itself be described as an archetypal person who is also a more complex eternal object than the simple eternal objects delineated above.

However, it should not be inferred that more basic eternal objects like the simple qualities Whitehead mentions above are ontologically prior to archetypal eternal objects like Saturn. Rather, because it is transtemporal, the formal realm of archetypes and eternal objects seems to paradoxically allow for the simultaneous validity of explanations that describe the archetypes as emergent from the simpler eternal objects, that describe the simpler eternal objects as qualities that emerge from complex archetypal persons, and that describe both the simpler eternal objects and the more complex archetypes as a priori. All three modes of explanation, though contradictory in the temporal realm, are not mutually exclusive because causation implies time. Understanding the relations of different classes of eternal objects to one another requires the employment of temporality as a metaphor because our language is inherently causal

However, as with all metaphor when pushed to an extreme, this one breaks down into paradox because, while there is no past, present, or future within the eternal archetypal realm, the quality of temporality is a manifestation of archetypes that do exist in that realm. Thus, like our mathematical descriptions of physical reality at orders of magnitude smaller than the Planck constant, our verbal tools, developed within temporality, are necessarily inadequate for describing the eternal realm. Similarly, as Le Grice notes, "Joseph Campbell recognized that the specific forms of the gods are expressions of more fundamental underlying principles."[29] This assertion by Campbell seems fundamentally to contradict Hillman's personified conception of the archetypes, which he describes as having "autonomy," "intentionality," and "a personal nature from the very beginning."[30] However, it seems likely that archetypes can be conceived fruitfully as both persons and impersonal forces, what Sri Aurobindo calls "Personalities and Powers of the dynamic Divine," for the transtemporal domain seems coextensive with the concept of nonduality found in many religious and spiritual traditions in which both poles of contradictory dualities can be seen as partial truths.[31] As Le Grice notes, by the time Jung wrote *Mysterium Coniunctionis*, he "had come to believe that behind the apparent multiplicity of the phenomenal world, psyche and cosmos form part of a unitary reality and rest upon a common 'transcendental background.' The archetypes, Jung proposes, are rooted in this underlying unity of the *unus mundus*."[32] Thus, it seems that, for the archetypal eternal objects, the distinctions between personal and impersonal, subjective and objective, and psyche and cosmos do not manifest until these pure potentialities are actualized in time, for the nondual, transcendental realm of archetypal eternal objects is prior to differentiation.

Synchronicity

However, more than this subtle, though important, classification of archetypes and eternal objects, the primary area where I differ from Griffin is in his characterization of synchronicity as "probably the weakest element in Jung's speculations," which I believe exhibits a fundamental misunderstanding of that concept.[33] According to Griffin, "by 'synchronicity' Jung means a meaningful but noncausal relation between two events," which is an understandable reading of Jung, but one that I would argue is fundamentally flawed since synchronicity seems to be a modern renomination of formal causality.[34] Griffin disputes what he perceives as Jung's

> exaggerated claim that the parapsychological experiments of J. B. Rhine "prove that the psyche at times functions outside of the spatio-temporal law of causality." Jung adds that, to make sense of these experiments, "we must face the fact that our world with its time, space, and causality, relates to another order of things lying behind or beneath it, in which neither 'here and there' nor 'earlier and later' are of importance."[35]

It seems that the conceptual issue at the root of Griffin's misinterpretation of synchronicity is a simple semantic error that is easily resolved in light of more recent developments in archetypal thought. In particular, it has become clear from the work of Tarnas and others that Jung generally uses the word "causality" to refer only to material and efficient causality, a conflation that is comprehensible in the context of the dominant scientific discourse of Jung's time, when this limiting of causality to only one half of Aristotle's causal quaternity was almost exclusively privileged.[36] Jung vaguely prefigures this insight when he writes in *Synchronicity* that "certain phenomena of simultaneity or synchronicity seem to be bound up with the archetypes," though these "certain phenomena" are never quite explicated in Jung's pioneering book.[37] Tarnas, with the perspective afforded by decades of collective work on these issues by archetypally-oriented thinkers, makes the connection more explicit:

> The occurrence of synchronicities is seen as permitting a continuing dialogue with the unconscious and with the larger whole of life while also calling forth an aesthetic and spiritual appreciation of life's powers of symbolically resonant complex patterning. . . . Although Jung himself did not explicitly describe this later stage in his principal monograph on synchronicity, it is evident from many scattered passages in his writings and from the recollections and memoirs of others that he both lived his life and conducted his clinical practice in a manner that entailed a constant attention to potentially meaningful synchronistic events that would then shape his understanding and actions. Jung saw nature and one's surrounding environment as a living matrix of potential synchronistic meaning that could illuminate the human sphere. He attended to sudden or unusual movements or appearances of animals, flocks of birds, the wind, storms, the suddenly louder lapping of the lake outside the window of his consulting room, and similar phenomena as possessing possible symbolic relevance for the parallel unfolding of interior psychic realities. . . . Central to Jung's understanding of such phenomena was his observation that the underlying meaning or formal factor that linked the synchronistic inner and outer events—the formal cause, in Aristotelian terms—was archetypal in nature.[38]

Thus, the "connecting principle" of synchronicity, which seems essentially to be a recasting of formal causality, results from the ingression into actuality of the archetypal class of eternal objects that exist outside, but are implicit in, the realm of temporality, spatial extension, and material-efficient causality. Based on Tarnas's insights, the Weberian "disenchantment of the world" was essentially constituted in the gradual repression of formal causation as a valid explanatory mode.[39] Thus, the reemergence of formal causation constitutes a gradual "re-enchantment of the world," which accounts for the numinosity that accompanies "synchronicities," moments in which formal causation seems to assert itself in one's psyche in the context of a disenchanted cosmological and metaphysical world view.[40] By contrast, in a world view that implicitly accepts formal causation, everything is potentially "enchanted."

Griffin seems to be correct when he writes:

> the only adjustment necessary is to broaden 'causality' so that it includes every degree of causal influence, not just deterministic forms, and nonlocal (noncontiguous) causal influence as well as local. And this is hardly revolutionary: thanks largely to quantum physics, this broadening of the scientific and philosophical notion of causality has already occurred.[41]

Thus, Griffin aptly recognizes that the way to clear up many of the seemingly insoluble problems of modern philosophy is to re-expand our definition of causality to encompass something approximating Aristotelian formal and final causation in a postmodern context. However, Griffin's dismissal of synchronicity is unnecessary and counterproductive to this project. Admittedly, Jung's articulation of synchronicity was somewhat inconsistent, a situation that Griffin recognizes when he writes that "the question of exactly how archetypes are to be understood is surrounded by controversy."[42] However, one might excuse Jung on this account because of what Jung himself refers to, in his foreword to Erich Neumann's *The Origins and History of Consciousness*, as his "pioneer" status. It was necessary for Jung to create a novel vocabulary in order to express his ideas, a task made all the more difficult by the subtle and complex nature of the archetypes. Furthermore, I agree with Tarnas's view that it was not so much that Jung was confused, but that his thought developed significantly over the course of his career:

> Jung's thought was extremely complex and in the course of his very long intellectually active life his conception of the archetypes went through a significant evolution. The conventional and still most widely known view of Jungian archetypes . . . was based on Jung's middle-period writings when his thought was still largely governed by Cartesian-Kantian philosophical assumptions concerning the nature of the psyche and its separation from the external world. In his later work, however, and particularly in relation to his study of synchronicities, Jung began to move toward a conception of

archetypes as autonomous patterns of meaning that appear to structure and inhere in both psyche and matter, thereby in effect dissolving the modern subject-object dichotomy.[43]

In light of Tarnas's work, the ultimate philosophical implication of Jungian synchronicity (as opposed to the psychological implications, which are not my primary concern) is that formal causality must be reinstated as a valid form of causation if our experience is to be understood in all of its multivalence. This insight is particularly relevant to the discipline of archetypal cosmology, as synchronicity is precisely the principle of complex causation by which the connection between the movement of the planets and the events in human life can best be understood. Indeed, Tarnas describes astrological correlations in *Cosmos and Psyche* as "one special, highly controversial class of synchronicities."[44] Whereas in a disenchanted cosmology that only acknowledges material and efficient causation, the meaningful correlation of the planets with events in human life is unintelligible, in a cosmology that accepts formal and final causation as valid, the correlation of "cosmos and psyche" becomes more readily intelligible.[45]

Panexperientialism

Nevertheless, based on his misunderstanding, Griffin views synchronicity and Whiteheadian panexperientialism as competitive explanations for the same phenomena. Griffin's explication of panexperientialism is illuminating:

"Memory" is the name we give to that peculiar relation we have to our own past. Although these past experiences now exist only as objects, we remember what they were in themselves, as subjects. A chimpanzee, most of us believe, has a similar relation to its past. The Whiteheadian suggestion is that this relationship applies analogously all the way down. A cell would therefore have some slight memory of what it experienced a few seconds earlier, and an electron an even slighter memory, perhaps going back no longer than a millionth of a second. The difference between subjects and objects, and therefore between psychic and physical energy, in other words, is the difference between present and past, and being known from within and being known from without. The difference is not ontological, but merely temporal and epistemological.[46]

Thus, panexperientialism is the recognition that all occasions, all matter and energy, have both exteriority and interiority and that these two realms are intimately connected. This recognition does not mean that rocks or electrons are conscious, but rather that inanimate matter always already contains within it the potentiality for consciousness, engaging in the subjective relationality, the *meaningfulness* that is the precondition for the emergence of consciousness. In Whitehead's view, all things contain within them the potential to communicate meaning to all other things, which, put in these terms, sounds suspiciously like a justification for formal causality and, thus, synchronicity. To use Whitehead's language, if all actual occasions participate in the concrescent (which roughly translates to "exhibiting increasing interconnection issuing into the emergence of novel entities") meaning of the world, and if all things contain within them the potentiality for consciousness and meaning that reaches its current known apex of development in the human mind, then the world can communicate its meaning through the human mind or, as Tarnas articulates it: "the human mind is ultimately the organ of the world's own process of self-revelation."[47]

Thus, when Griffin writes that, "although Jung himself sought to interpret . . . [parapsychological] phenomena as further examples of synchronicity, they can be interpreted better in terms of process theology's panexperientialism," he is unintelligibly privileging one concept over another concept when the two concepts are actually mutually necessary and reinforcing, like two distinct mathematical formulae that can be shown to be exactly equivalent through a series of transformations.[48] Panexperientialism and synchronicity, understood in its most mature formulation, are complementary concepts that imply one another beneath the umbrella of formal causality. That is, Whitehead's idea that all matter and energy have interiority is a precondition for the interior connectivity implied by Jung's idea that inanimate objects and human consciousness can meaningfully participate in the same archetypes. These concepts have clear application in archetypal cosmology, which posits a significant interior correlation between planets

Beyond a Disenchanted Cosmology

and human experience for, as Jung writes in *Synchronicity*: "the meaningful coincidence we are looking for is immediately apparent in astrology."[49] Thus, by synthesizing the work of Jung and Whitehead, it can be said that the archetypal eternal objects are potentialities of meaning that exist prior to spatio-temporality, informing the meaning of the world that we find ourselves in through both synchronicity and panexperientialism, both of which posit an intimate connection between interiority and exteriority, subject and object, psyche and cosmos through their mutual participation in formal causation.

In fact, there appears to be a fractal structure embedded in this formulation, like a multi-layered pun communicated to us by the structure of being or the dynamics of process: synchronicity is the operation whereby two seemingly unconnected events, such as an external occurrence and an internal state, two simultaneous external events, or two events separated in time, are seen to be manifestations of the same archetypal complex. Analogously, or perhaps fractally, synchronicity and panexperientialism—two seemingly unconnected concepts—are both inflections of formal causality because they both deal with the interiority of things. Thus, the same structure of meaning is operative on different levels of significance, which seems to demonstrate the fractal property of self-similarity across scale: on the one hand, the level of specific individual occasions of synchronicity that connect two seemingly unconnected events and, on the other hand, the more abstract conceptual level where the two seemingly unconnected concepts of synchronicity and panexperientialism are shown to be related by their mutual participation in formal causation. Perhaps this fractal pattern can be extended to a third level to encompass the mutual participation by Whitehead and Jung in the conception of a new world view.

Notes

1. David Ray Griffin, *Archetypal Process: Self and Divine in Whitehead, Jung, and Hill-man.* (Evanston, IL: Northwestern University Press, 1989), vii.

2. Griffin, *Archetypal Process*, 2. Griffin defines "process theology" as "the movement originating with Alfred North Whitehead" (*Archetypal Process*, vii), though I believe that "process philosophy" is more accurate as Whitehead saw himself primarily as a philosopher, not a as a theologian.

3. Griffin, *Archetypal Process*, 13.

4. Keiron Le Grice, *The Archetypal Cosmos: Rediscovering the Gods in Myth, Science and Astrology* (Edinburgh: Floris Books, 2010), 160.

5. Griffin, *Archetypal Process*, 16.

6. Richard Tarnas, *Cosmos and Psyche: Intimations of a New World View* (New York: Viking, 2006), 1.

7. Tarnas, *Cosmos and Psyche*, 16.

8. Richard Tarnas, *The Passion of the Western Mind: Understanding the Ideas That Have Shaped Our World View* (New York: Ballantine, 1991), 432.

9. Tarnas, *Cosmos and Psyche*, 84.

10. Richard Tarnas, "Archetypal Cosmology: Past and Present," *The Mountain Astrologer* (April/May 2011). According to the syllabus for that course, the conference on which the book *Archetypal Process* is based constitutes "perhaps the fullest academic anticipation of the concerns and themes that later came to inspire the transdisci-plinary focus of the PCC program," of which Tarnas was the founding director, and which has been one of the primary incubators for archetypal cosmology.

11. Griffin, *Archetypal Process*, 11.

12. Griffin, *Archetypal Process*, 27.

13. See Griffin, *Archetypal Process*, 26–39. Moreover, James Hillman, in his essay in the same volume, "Back to Beyond," does in fact write rather extensively about astrology in the three final sections of the essay, beginning with the section entitled "Psycho-logical Cosmology," though, as with most of Hillman's writing on this subject, it is difficult to pin down what he considers to be the precise relationship of the planets to the gods with which they are associated. Unfortunately, Griffin does not address this central aspect of Hillman's essay in his response. For Le Grice's discussion of this topic, see Keiron Le Grice, "Astrology and the Modern Western World View," in *Beyond a Disenchanted Cosmology, Archai: The Journal of Archetypal Cosmology*, Issue 3 (San Francisco: Archai Press, 2011), 32–37.

14. Philip P. Weiner, *Dictionary of the History of Ideas* (New York: Scribner's, 1973), 272.

15. For an in-depth discussion of causality in relation to archetypal cosmology, see Le Grice, *Archetypal Cosmos*, 113–116.

16. Griffin, *Archetypal Process*, 10–11.

17. Griffin, *Archetypal Process*, 11.

18. Carl Gustav Jung, *Synchronicity: An Acausal Connecting Principle* (Princeton, NJ: Princeton University Press, Bollingen Series, 1973), 20.

19. Griffin, *Archetypal Process*, 11.

20. Alfred North Whitehead, *Process and Reality* (New York: The Free Press, 1978), 87.

21. The discussion in this paragraph is based on a conversation I had with Whitehead scholar Eric Weiss

22. Whitehead, *Process and Reality*, 61.

23. Whitehead, *Process and Reality*, 44.

24. James Hillman, *Re-Visioning Psychology* (New York: Harper, 1975), xx.

25. Hillman, *Re-Visioning Psychology*, 22.

26. Hillman, *Re-Visioning Psychology*, xix.

27. Le Grice, *Archetypal Cosmos*, 68.

28. Whitehead, *Process and Reality*, 46.

29. Le Grice, *Archetypal Cosmos*, 69.

30. Hillman, *Re-Visioning Psychology*, 22.

31. Le Grice, *Archetypal Cosmos*, 179.

32. Le Grice, *Archetypal Cosmos*, 171.

33. Griffin, *Archetypal Process*, 27.

34. Griffin, *Archetypal Process*, 27.

35. Griffin, *Archetypal Process*, 27–28.

36. See Sonu Shamdasani, *Jung and the Making of Modern Psychology: The Dream of a Science* (Cambridge University Press, 2004) for examples of popular theories in the late nineteenth century that presented variants on the notion of formal causality. As is almost always the case, any privileging operation, like that of material-efficient causation over formal-final causation, is always already deconstructed by the inevitable examples that contradict that privileging.

37. Jung,, *Synchronicity: An Acausal Connecting Principle*, 21.

38. Tarnas, *Cosmos and Psyche*, 56–57.

39. Tarnas, *Passion of the Western Mind*, 412.

40. See Morris Berman's *The Re-Enchantment of the World* (Ithaca, NY: Cornell University Press, 1981).

41. Griffin, *Archetypal Process*, 31.

42. Griffin, *Archetypal Process*, 39.

43. Tarnas, *Passion of the Western Mind*, 425.

44. Tarnas, *Cosmos and Psyche*, 61.

45. Tarnas, *Cosmos and Psyche*, 50–60.

46. Griffin, *Archetypal Process*, 23.

47. Tarnas, *Passion of the Western Mind*, 434. According to Whitehead, "the process, or concrescence, of any one actual entity involves the other actual entities among its

components. In this way, the obvious solidarity of the world receives its explanation" (*Process and Reality,* 7).

48. Griffin, *Archetypal Process*, 32.

49. Jung, *Synchronicity: An Acausal Connecting Principle*, 38. Although this statement seems fairly clear, Jung writes a mere six pages later: "One would therefore do well not to regard the results of astrological observation as synchronistic phenomena, but to take them as possibly causal in origin," (44–45) which seems to contradict the statement made in the main text. Thus, Jung seems to have remained undecided about whether astrological correlations were instances of synchronicity or of efficient causation, whereas the consensus within archetypal astrology seems to be that synchronicity is the mode of causation whereby the movements of the planets can be meaningfully correlated with the events in human experience.

Bibliography

Berman, Morris. *The Re-Enchantment of the World*. Ithaca, NY: Cornell University Press, 1981.

Griffin, David Ray. *Archetypal Process: Self and Divine in Whitehead, Jung, and Hillman*. Evanston, IL: Northwestern University Press, 1989.

Hillman, James. *Re-Visioning Psychology*. New York: Harper, 1975.

Jung, C. G. *Synchronicity: An Acausal Connecting Principle*. Translated by R. F. C. Hull. Princeton, NJ: Princeton University Press, Bollingen Series, 1973.

Le Grice, Keiron. *The Archetypal Cosmos: Rediscovering the Gods in Myth, Science and Astrology*. Edinburgh: Floris Books, 2010.

———. "Astrology and the Modern Western World View." In *Beyond a Disenchanted Cosmology. Archai: The Journal of Archetypal Cosmology*, Issue 3 (Fall 2011). San Francisco: Archai Press, 2011, 32–33.

Shamdasani, Sonu. *Jung and the Making of Modern Psychology: The Dream of a Science*. Cambridge: Cambridge University Press, 2004.

Tarnas, Richard. "Archetypal Cosmology: Past and Present," *The Mountain Astrologer*, no. 157, (June/July 2011): 65-59.

———. *Cosmos and Psyche: Intimations of a New World View*. New York: Viking, 2006.

———. *The Passion of the Western Mind: Understanding the Ideas That Have Shaped Our orld View.* New York: Ballantine Books, 1993.

Weiner, Phillip P. *Dictionary of the History of Ideas.* New York: Scribner's, 1973.

Whitehead, Alfred North. *Process and Reality: An Essay In Cosmology: Gifford Lectures Delivered In the University of Edinburgh During the Session 1927–1928.* Corrected Edition. Edited by David Ray Griffin and Donald W. Sherburne. New York: The Free Press, 1978.

Reflections on Archetypal Astrology and the Evolution of Consciousness

Sean M. Kelly

Introduction[1]

Richard Tarnas has now written two extraordinary books, both of them as beautiful as they are profound. The first, *The Passion of the Western Mind*, is more than a history of ideas, since it employs as it extends a psycho-philosophy of history or a theory of the evolution of consciousness whose perspective follows broadly Hegelian and Jungian lines. The deep structure of this perspective has been described by M. H. Abrams as a "circuitous journey" and corresponds to a particular (and, as I discuss in *Coming Home*, essentially Christian in origins) inflection of what Joseph Campbell called the "monomyth" or hero's journey with its movement out of an original, pre-egoic identity, through an initiatory encounter with death, to a new, more conscious and integrated identity. The second book, *Cosmos and Psyche*, presents a larger body of evidence and a sustained argument for the value and implications of archetypal astrology for an understanding of the complex movements of culture, with a focus on the birth of the modern and the "intimations," as he puts it in the subtitle, of a new worldview. While the general subject of astrology receives a brief historical treatment in the *Passion*, it is otherwise not explicitly engaged, though in retrospect (that is, after reading the second book) one can see how the archetypal-astrological perspective has informed his presentation of the principal dynamic factors involved in the evolution of the western mind. *Cosmos and Psyche*, for its part, though it retains some of the core elements of the Hegelian and Jungian orientations, gives most attention to several planetary cycles which reveal the periodic surges of the corresponding archetypal, cultural, and historical streams or movements under consideration. The relationship between the two main hermeneutical perspectives—the Hegelian-Jungian and the astrological—though in some sense overlapping and clearly in dialogue with the notion of the "archetypal-astrological"—has yet to be explicitly articulated. At the very least, more can be said as to just how, and with what consequences, the kind of integration that Tarnas has effected can be accomplished. While much of what I will propose is already implicit in Tarnas's written work, or has arisen in dialogue with him, I hope that the following reflections might indicate some fruitful directions for further discussion.

Terra Stella Nobilis

Tarnas's integration of the astrological perspective with that of the evolution of consciousness involves a recognition of the living Earth as the focal point or axis around which the planets (and stars) make their rounds.[2] Despite what might appear to the more literal-minded as a reversion to premodern geocentrism, this idea actually deepens Kant's insight that his proposal for a "transcendental idealism" constituted a second Copernican Revolution, only here, with a third revolution, it is not the isolated human subject which provides the epistemological meta-point of view, but the Earth as a whole (though still mediated, of course, through human knowing). Such a recognition allows us to honor to the fullest extent Kepler's insight—*terra stella nobilis*—that the Earth is a noble star, and in one sense the noblest among our solar system, in that it is through human consciousness as it has evolved on Earth that the new structure of the cosmos is revealed. Such an honoring would, finally, be more consistent with the simultaneously organicist and anthropocosmic orientations of Hegel and Schelling (and other *Naturphilosophen* influenced by them). For orientations such as these, the Earth (that is, its life processes, and especially as these come to self-consciousness in humans) represents a more concrete, because more complexly organized, and therefore more ensouled being than the other planets and stars.[3] This view, it should be stressed, signals a marked reversal of the dominant view of the ancients, for whom the planets and stars were considered more perfect beings than the Earth (and its four elements).

From the point of view of traditional astrology, the movements of the planets are without a proper history, in the sense of a directed narrative with beginning, middle, and end (this is less true of the gods and goddesses to which, in the mythic imagination, the planets correspond). Their motions are described as the more or less mechanically (and to the ancients, eternally) recurring cycles we see plotted out in the ephemeris. As the ancients conceived of it, only the "sublunary" realm of the Earth and its elements are given over to flux and becoming. The stars and planets were thought to consist of the immutable fifth element (the ether) or to be associated with imperishable crystalline spheres. Both ancient and contemporary astrology do indeed see the "influence" of these cycles on, or their correlation with, human affairs. It is my understanding, however—and Tarnas demonstrates this in practice—that these cycles, *by themselves*, cannot account for the overarching trajectory of the history or evolution of consciousness or of culture.

By contrast, through his Romantic-archetypal reading of astrology—and his astrological reading of archetypes—Tarnas was already able toward the end of *Passion* to characterize this evolution in terms of a struggle or dialectic between Uranus (or Prometheus) and Saturn, or at a deeper level, between the archetypal masculine and feminine (which overlaps with the solar and lunar archetypes). But to give an account of the actual history of the human project, especially as this has unfolded over the historical period and into that of the Planetary Era—to give an account, that is, of the evolution of consciousness—one needs to attend to the essentially unrepeatable, non-reversible, and non-cyclic (if even only as a spiral) succession of events and mythemes that have in fact marked the history in question. The story of the Earth, in other words, is not only illumined by, but illuminates or *provides the point of view* from which to make sense of the archetypal dance of the planets.

A key consideration here is the discovery of the most studied trans-saturnian planets—Uranus, Neptune, and Pluto—which play such a central role in the archetypal-astrological perspective. Unlike the five visible planets plus the Sun and Moon, whose archetypal meanings seem to have been given all at once, or in any case are essentially fixed from the earliest records to the present, the meanings of Uranus, Neptune, and Pluto were initially determined through extensive study of their positions in natal charts and in mundane astrology (that is, the correlation of their positions with world events). Tarnas has added to this a kind of synthetic intuition of the overall configuration of the world soul (*anima mundi*) at the time of their discovery. Uranus, for example, which Tarnas has identified with the mythic Prometheus, reveals itself (among other things) through the breaking into history of the principle of freedom in its distinctively modern form, with its ties to the great political revolutions and the industrial revolution of the late eighteenth-century. It is certainly significant that, after the fact, one can point to earlier and subsequent moments in the Uranus cycle (in its quadrature alignments with other planets, especially Jupiter, Saturn, Neptune, and Pluto) which also manifest a rising to the surface of the promethean impulse. Still, at least in the case of Uranus—the first of the trans-saturnians, and thus the point of breakthrough out of the premodern planetary cosmic picture where the Earth and human life is in some sense subject to the "influences" of the endlessly repeating cycles of the visible planets—there is something unrepeatable, qualitatively unique, and one could even say paradigmatic, about the historical moment corresponding to the discovery of this planet. It is through the spirit of this unique moment that the meaning of the entire sequence of prior and subsequent uranian moments is revealed. But again not in practice, only retrospectively.

History, Evolution, and the Great Code

In this case, what is revealed is that *freedom has a history*. Again, if we confine ourselves to the quadrature alignments of Uranus with Saturn or Neptune or Pluto, we will see the rhythmic pulses of the drive for freedom and discovery, but we will not discern the entelechy or the organic patterning of the sequence of emancipatory pushes. An analogy might help here. It is well known that human development has a certain periodicity, with the seven and a half year cycle being especially marked (this corresponds, of course, to the four phases of the twenty-eight year Saturn cycle). At the same time, however, there is an irreducible difference between the kind of transition that tends to occur around age

7 as compared with the transition of puberty (around age 14) or the beginning of full maturity around 28–30. The concrete meaning of each transition cannot be derived from the nature of the square (age 7) or opposition (14) or new square (21) but only from the context of the concrete unfolding of the organism in its passage from birth to death. This becomes even more obvious when the first Saturn-return (centered around age 30) is compared with the second (around age 60), both of which are formally identical in terms of aspect. While it is possible to formulate a philosophy of history or theory of the evolution of consciousness according to a purely cyclic-organismic pattern (as we see with Vico, Spengler, and to a certain extent even with Toynbee), it would seem that an additional perspective or analogy is required to make sense of the "arrow of time" with its associated qualities of radical specificity, irrepeatability, and genuine novelty.

Tarnas's approach to the idea of evolution, of course, is more in line with Teilhard and the Hegelian-Jungian perspective, which is not only organismic but is stretched between an Alpha and an Omega (however these are interpreted), with each station in between representing a true emergence of greater depth, complexity, and consciousness (though in the more Jungian and Romantic inflection favored by Tarnas, and myself, the middle, "disenchanted" phase is seen as involving a particular kind of loss as well). In any case, the appeal to the idea of evolution must involve a contextualization of the merely cyclic movements of the planets to the life of the greater organism of the evolving cosmos and of the Earth, both of which have a definite history with a beginning, an unfolding and essentially unforeseeable drama, and presumably some kind of end.

Arthur Lovejoy made famous the understanding of evolution as the "temporalization of the Great Chain of Being." This understanding can be seen as a specific corollary of the more general proposal—as argued by Karl Löwith, for instance—that the modern worldview (which birthed the idea of evolution, and before that the ideal of progress) must be understood as the secularization, and this as an organic metamorphosis, of the formerly dominant Christian worldview. The spirit of this proposal is perhaps best summed up in Blake's lapidary pronouncement that "The Old and New Testaments are the Great Code of Art." In this light we can see with greater clarity just how the emergence of the modern and, more importantly, the Planetary, Era is directly linked to what McNeill termed the "rise of the West," whose guiding mythos or symbolic matrix has been the Biblical, and specifically Christian view of time and history as directed or at least as irrevocably marked by a singular Event, the meaning of which can of course be interpreted from a number of perspectives—from the literalist/fundamentalist to the secular/ humanist to the more interesting possibilities offered by speculative theology, philosophy, and psychology (as with Teilhard, Hegel, or Jung). Without such explicitly speculative considerations, an astrological perspective can illuminate and deepen one's understanding of any particular kind of *moment* or *series* of moments of the historical process (with each occurrence of the quadrature aspects in the cycles of the outer planets, for instance), but it cannot, by itself, provide a comprehensive or fully coherent account of the *meaning* of history or the evolution of consciousness. That is, unless, as we see with Tarnas's Romantic-archetypal approach, the astrological perspective truly comes to terms with, and explicitly honors, its evolutionary geocentrism. This means acknowledging the need to complement, or rather *complete, both the astrological and archetypal perspectives with the planetary/historical*—again, not only by seeing how the former is played out in the latter, but also how (in what specific ways, according to which particular embodiment of the fundamental pattern) the movements of time, evolution, and planetary history have actually played themselves out (which is exactly what Tarnas succeeds in doing in *Cosmos and Psyche*).

From the *Anima Mundi* to the *Weltgeist*

A possible critique of *Cosmos and Psyche* concerns its focus on Western culture and historical events.[4] If planetary alignments indicate a universal field of archetypal meaning, so the critique would run, one should expect to see this meaning equally reflected in the lives and happenings of people from around the globe at whatever time is being selected. Tarnas treats five periods that do seem to satisfy this requirement in a striking way (the five are: c. 570 BCE, the 1640s CE, the 1840s, c. 1900, and the 1960s). The first has to do with the only triple conjunction of Pluto, Neptune, and Uranus in historical times, from the 580s to the 560s BCE. This was the high point of the first Axial

Period, which witnessed the more or less simultaneous emergence of many of the world's major religious and philosophical traditions, including the Presocratics in Greece, Lao Tzu in China, Zoroaster in Persia, the Buddha in India, and the major Jewish prophets in the Near East. Closer to our own times, there is the Uranus-Pluto conjunction of 1968, which coincided with a wave of student protests and a generalized mobilization of the youth in major cities around the world, including Berkeley, Chicago, New York City, Mexico City, Paris, Prague, and Beijing (a similar wave of revolutionary protests swept across Europe in 1848 with the previous conjunction).

From a planetary perspective, however, it would seem that one or only a few events can be singled out as the most emblematic manifestations of the alignment in question. A case in point concerns the 9/11 attacks of 2001. It is pointed out by astrologers that the attacks occurred during a Pluto-Saturn opposition. It is also often noted, as it is by Tarnas in his consideration of the Pluto-Saturn cycle, that the prior quadrature alignments of this cycle coincide with the outbreak or intensification of major wars or armed conflicts, as with the onset of the two World Wars and the Vietnam War earlier in the last century. It might legitimately be asked, however, why other *World* Wars, or an event such as 9/11 with clear world-historical consequences, didn't happen during every conjunction, square, or opposition of these two planets? The World Wars seem to stand out as events without clear parallel in terms of numbers of dead and injured and for the sheer massiveness of people and energies involved. As for the 9/11 incident, though dwarfed by the World Wars, it nevertheless marked the beginning of the global "war on terror" that has dominated so much of the political landscape since that sorry day. Still, these three events correspond to only three among twelve quadrature alignments during the three turns of the Pluto-Saturn cycle from the beginning of the twentieth century to our own times. Arguably, none of the other nine alignments involved events that match the World Wars or 9/11 in scope or consequence (though they can of course be seen as leading to or following from, the more emblematic events). Clearly, therefore, the cycle by itself is incapable of accounting for the radical particularity of the events in question. Perhaps, as with the discovery of Uranus around the time of the French and American Revolutions, the discovery of Pluto in the early part of the twentieth century could be adduced to qualify the momentousness of this particular iteration of the cycle. But this, as I see it, is to appeal to the unique, singular, or irreversible character of our historical unfolding, and not to the cycle itself.

In fact, it is only within the context of the overall trajectory of the evolution of consciousness that the principle of selection becomes apparent. More particularly, it is necessary to discern the movements of what Hegel called the *Weltgeist*, the world spirit, which can be thought of as expressing the leading edge or focus of intentionality within the overall trajectory. Once we grant the unique role of the West—Europe to begin with, then especially the United States—in bringing about the birth of the Planetary Era through the world-transforming combination of the modern secular ideals of progress and freedom with the darker colonial aspirations of the great nation states (some of them destined to become superpowers) and the overwhelming force of the industrial and continuing technological revolutions, one might be justified in looking to the West for the signal events corresponding to any given major planetary alignment. While the world soul is, as it were, evenly spread across the entire globe, the world spirit—which is the spirit of history itself—would seem to concentrate itself wherever those events are transpiring which will have the greatest consequences for the further unfolding of the overall trajectory (from a Jungian perspective, one could consider the *Weltgeist* to function as a kind of planetary ego—or perhaps what Jungians refer to as the ego-Self axis—while the *anima mundi* would correspond to the collective unconscious).[5] In this way, not only would one not expect to see equally powerful or significant expressions of a given planetary combination with each of its quadrature alignments, but it could make sense to focus on the 9/11 event, for instance, as the most significant manifestation of that particular Pluto-Saturn alignment (rather than on some other conflict elsewhere in the world that did not receive as much attention). The main point, once again, is that the archetypal-astrological perspective is most coherent and compelling when explicitly engaged with the larger and more determinate context of the evolution of consciousness.

Freedom and the *Hieros Gamos*

According to Hegel, the goal of world history, the deepest passion or longing of the world spirit, is the actualization of the principle of freedom.[6] We have already seen that a merely (that is, non-Romantic-archetypal) astrological perspective on the Uranus-Pluto cycle, though it might serve to identify a series of historical moments of manifest freedom (and point to resonances between such moments), cannot by itself reveal the overarching pattern or trajectory of history or the evolution of consciousness. Seen within the context of this trajectory, a few critical moments do serve to throw this pattern or trajectory into relief. While the early Greek ideals of autonomy and democracy are clearly relevant to the story, Hegel claims that it is Christianity (with its unique stress on the absolute value of the individual) that introduces the principle of universal freedom into the world-historical process. It is, in effect, by tracking the working out of this principle over the centuries that the overall pattern or trajectory is revealed. Following the implantation of the principle in mythic form with the teaching of the gospels, two of the most significant subsequent events are the Protestant Reformation—which absolutizes the freedom of individual conscience—and the American and French Revolutions, which establish the now secularized ideal of freedom as the universal standard of the modern state. Needless to say, none of these events brought about the full realization of the principle of freedom. The main point, however, is that the movement of history over the last two millennia—and in this case the particular movement of the actualization of freedom—manifests a definite trajectory, marked by a number of critical phase-shifts, the specific character and telos of which is invisible to a perspective that restricts itself to the ever-recurring cycles of traditional astrology.

While Tarnas adopts the general three-phase structure of the Hegelian dialectic for his big picture of the history of the Western mind, and more generally for the evolution of human consciousness, instead of privileging the actualization of the principle of freedom, he follows Jung in characterizing the goal or telos in more symbolic or imaginal terms as a *coniunctio oppositorum* or sacred marriage (*hieros gamos*). This characterization is of course fully consistent with the third moment of Hegel's dialectic, which can be summed up with the notion of the *identity of identity and difference* (and it is in fact by the standard of such a complex identity that one can evaluate the concreteness or actuality of the principle of freedom). Rather than such concepts as identity and difference, or universality and particularity, however, Tarnas's preferred terms for the *coniunctio* or marriage are *psyche* and *cosmos* and, as we have seen, at a more archetypal level, the *masculine* and the *feminine*. In keeping with the Jungian view of development, the (re)union of these terms represents the overcoming of their alienation, especially during the modern period following the Copernican Revolution and the rise of the Cartesian-Newtonian paradigm. What I am suggesting here is that the sacred marriage as envisioned by Tarnas calls for a more explicit celebration of a second *coniunctio*, or *complexio oppositorum*, as Jung also defined the Self (and which he conceived as both the guiding spirit and goal of the individuation process or the evolution of consciousness): in this case the joining or weaving together (*com-plexere*) of the more symbolical/archetypal and polycentric perspective of astrology with the more conceptual/dialectical and monocentric perspective of the evolution of consciousness (as exemplified by Hegel, Teilhard, Steiner, and Aurobindo).

Though the full import of the role of astrology would have to await the publication of *Cosmos and Psyche*, the union (or at least the passionate engagement) of these two perspectives is already in evidence in *The Passion of the Western Mind*, where both Hegel and Jung play central hermeneutical roles. At about the same time as the appearance of *Passion*, though neither of us knew of the other's work, I argued a more restricted and technical case for such a union in my *Individuation and the Absolute*, where I demonstrated that Jung's concept of the Self is both structurally and functionally equivalent to Hegel's concept of Spirit (*Geist*) or the Absolute. Because the union is complex, however, I concluded that the terms in relation will maintain a certain creative tension—it is a matter, after all, of the identity of identity *and* difference. The same holds true for the present discussion. We have seen repeatedly how the cyclical perspective of astrology is, taken by itself, blind to the arrow of time. The conceptual-dialectical perspective, for its part, can be somewhat tone-deaf to the music of the spheres, to the subtler symbolic-imaginal resonances in the sounding of a given historical moment or between related moments. Each in their difference, in other words, is well served by the other.

Beyond a Disenchanted Cosmology

Conclusion: The Spiral of Evolution

It is a truism among those given to the spirit of mediation that the apparent contradiction between the arrow of time and the ever-recurring cycle is resolved in the image of the spiral. The challenge has been to describe exactly what the spiral of history or evolution would look like in the details. In *Coming Home: The Birth and Transformation of the Planetary Era*, I have outlined an approach to the last two millennia, at least, where such a spiral appears to be in evidence.[7] To summarize: What I found was that the three-phase structure of the overall trajectory seems to be fractally repeated, at least once between the first and second phases of the larger arc, and several times between the second and third. The three main phases are: 1. identity: The Biblical worldview as Great Code; 2. difference: The Modern worldview (as secularization of the Great Code); and 3. an accelerated movement toward the identity of identity and difference: The Planetary worldview (a new identity in the making). The first turn of the spiral is traced by the movement from the life of the early, mythically embedded, Christian community, through the differentiation of consciousness effected by medieval scholasticism and the establishment of the worldly power of Christendom, to the birth of modernity with the Renaissance, the Copernican Revolution, and the Protestant Reformation, which together, despite their negation of the preceding medieval worldview, represent the forging of a new, more complex identity. This new identity is itself negated with the Enlightenment and the emergence of the fully secularized (and materialistically oriented) Cartesian-Newtonian paradigm, which in turn generates a second fractal third moment in the counter-cultural response of Romanticism and speculative Idealism (culminating in Hegel).

I make the case of there having been three more ever-tightening turns of the spiral from then to the present. Despite the revolutionary developments in physics, psychology, spirituality, and the arts surrounding the threshold of the twentieth-century, the last two centuries have been dominated by what Van Baumer calls the New Enlightenment (which includes such movements as positivism, Marxism, Freudianism, and a general faith in the power of economic and technological progress). Without actually overturning the dominant trend, the sixties Counterculture succeeded in constituting a distinct new identity while taking up, however unwittingly, many of the themes and much of the spirit of the earlier countercultural surge of the Romantic-Idealist era (especially with its organicist, participatory, and enchanted view of the cosmos). As we know, the Counterculture was succeeded by a conservative entrenchment in politics and in culture generally. At least in the United States, one can discern a third turn of the spiral with the movement through the more optimistic and spiritually open years of the Clinton administration to the current phase dominated by the neo-conservative agenda. Encouraged by Tarnas's reading of approaching world transits, I have risked the prediction that we are on the threshold of another countercultural upsurge—though of course, with mounting tensions and deepening complexity, everything is increasingly uncertain.

One of the more compelling indicators, which confirms the rate at which the spiral is tightening, are the immanent world transits involving Pluto, Uranus, and Saturn, the configuration of which shows a clear resonance with the sixties Counterculture (and earlier with the time of the French Revolution and Terror immediately preceding the birth of the Romantic-Idealist era). Despite these illuminating correspondences, however, I am left with the following observations with respect to the union of the astrological with the Hegelian and Jungian perspectives: Even in spiral form, the overall trajectory points to a kind of Omega point or Singularity. What are the implications for our collective human relationship to the planetary archetypes as we draw ever nearer to this Singularity, or perhaps even pass through it to whatever lies on the other side? Early Christians, both Gnostic and orthodox, imagined that the singular salvific Event by which their time—and all time—was defined had the potential of freeing them from the fatalism of the stars. Two millennia later, we have the potential, at least, to feel more at home in the cosmos. Still, at the time of writing, a sixty-five million year geological era (the Cenozoic) is coming quickly to an end through the human created Sixth Mass Extinction, the very biosphere as we have always known it is threatened with collapse, the resource base of the modern era is being rapidly depleted, and there are billions of human lives trapped in desperate suffering. The goal of fully actualized freedom on a planetary scale seems impossibly distant. While the archetypal-astrological perspective or that of the spiral of evolution I have traced in *Coming Home* may illuminate the path immediately ahead, the Earth itself has become a "wanderer" (*planētēs*) in a sense scarcely imaginable by the ancient astrologers. Whether here on Earth or directed to the heavens, we only see as in a glass, darkly . . .

Notes

1. This article has been adapted from a chapter of Sean Kelly's *Coming Home: The Birth and Transformation of the Planetary Era* (Hudson, NY: Lindisfarne Books, 2010).

2. See Tarnas, *Cosmos and Psyche: Intimations of New World View* (New York: Viking Books, 2006), 489 and 492.

3. A fascinating exception here is G. T. Fechner. Though a student of a student of Lorenz Oken (who was a student of Schelling), and sharing many other elements in common, Fechner proposes that, just as our individual souls effect a conscious merger with the *anima mundi* after death, so the *anima mundi* participates in the larger soul of our Sun, which participates in the soul of the galaxy, etc. In contrast to Schelling and Hegel, in other words, for Fechner human consciousness is a transitional (rather than the central) middle term in the cosmic evolutionary scheme.

4. For his position on this matter, see Tarnas, *Cosmos and Psyche*, 137.

5. Hegel remarked on the westward movement of the *Weltgeist*. Giving the lie to those who claim that he saw world history ending with the Prussian state, he suggested that the focal point of the *Weltgeist* would likely move out of Europe to the United States, and this more than a century before the latter's rise to the status of superpower.

6. As Richard Tarnas reminded me, Rudolph Steiner would say both freedom and love, which would actually be consistent with the more Romantic phase of Hegel's early thought.

7. See Kelly, *Coming Home*.

Bibliography

Kelly, Sean. *Coming Home: The Birth and Transformation of the Planetary Era*. Hudson, NY: Lindisfarne Books, 2010.

Tarnas, Richard. *Cosmos and Psyche: Intimations of New World View*. New York: Viking Books, 2006.

Archetypal Analysis of Culture and History

Seasons of Agony and Grace

An Archetypal History of New England Puritanism
(Part Three): The Great Awakening

Rod O'Neal

This is the third essay in a series that presents an archetypal history of three major historical periods readily seen in the history of New England Puritanism. Part one, published in the first issue of *Archai*, examined the archetypal themes presented by foundational events leading up to and including the emergence of the Puritan movement in England in 1566–1567. Those themes, you will recall, presented a striking pattern of archetypal correlations with axial alignments of two major planetary cycles, Uranus-Neptune and Saturn-Pluto. Part two, published in the second issue of *Archai*, explored the next major historical period of the Puritan Movement, the mid-seventeenth century (roughly 1630–1660), when Puritans in both England and New England were involved in the expression, elaboration, and evolution of their religious utopian vision through a series of crises and controversies of historical consequence. These developments again displayed striking archetypal correlations with the world transit alignments of that period, especially the hard-aspect alignments between Saturn and the three outer planets—Uranus, Neptune, and Pluto—all taking place in the context of conjunction and opposition alignments in the longer-term Uranus-Neptune, Uranus-Pluto, and Neptune-Pluto cycles.

This third part of the study considers the period of the Great Awakening, in the 1730s and 1740s, when a dramatic and widespread wave of religious conversion experiences and spiritual awakenings occurred within the Puritan movement, with dramatically significant historical consequences. As in the previous two periods, these seminal events correlated especially with an axial alignment of Uranus-Neptune. And once again, a great many of the most intense, controversial, widespread, and enduring developments of the Great Awakening were in synchrony with the Saturn-Pluto square alignment of that period and were saturated with that complex's archetypal themes.

Awaiting a Season of Grace

For most of a century after its founding, New England prepared and waited for a mighty wave of church renewal that would swell the ranks of the elect, fill church pews, and confirm the religious vision of the New England Way. The innovations of the Half-Way Covenant did much to increase church participation and the Mass Covenant Renewals of the Reforming Synod bolstered piety for a while, but neither contributed significantly to the primary purpose of the Puritan errand into the wilderness. Periodically, individual congregations would experience a brief wave of revival, but most were short-lived and few rippled far beyond the nearest neighbors.

One of the most successful preachers to generate local revival waves was Solomon Stoddard, minister of Northampton and grandfather of Jonathan Edwards. Along with many other Puritan ministers, Stoddard advocated a theory of church revival that involved active roles for individuals, congregations, and preachers.[1] Revival only occurred, of course, by the will of God, who poured out his Spirit to dispense grace. But in addition to the many ways each person could prepare his or her soul for accepting the experience of grace, individuals and entire congregations could "fetch down" revival through prayer; and when spiritual conditions were just right, effective sermons could "preach up" revivals by catalyzing conversion experiences.[2]

The essential element, however, was the Spirit, whose presence was thought to ebb and flow according to a rhythm established by and known only to God.[3] Puritan ministers, therefore, were ever vigilant for signs of the arrival of particular seasons of awakening when God chose to pour out his Spirit in extraordinary abundance. Such "seasons of grace"—as they were commonly referred to in Puritan literature of the period—were modeled after the outpouring of the Spirit at Pentecost to the apostles as described in the New Testament, when "there appeared to them tongues as of fire, distributed and resting on each one the them. And they were all filled with the Holy Spirit" (Acts 2:3–4).[4] The Pentecost event was the fulfillment of the promise God spoke through the prophet Joel: "And in the last days it shall be, God declares, that I will pour out my Spirit upon all flesh" (Joel 2:28).[5] Although people could pray to fetch it down and ministers could preach it up, the time and place of these extraordinary outpourings of the Spirit could not be known. According to these evangelical Puritan divines, other than Pentecost and the early Reformation of the sixteenth century when thousands of people were converted to the true church, no other major season of grace had been witnessed—until the Great Awakening.

Most historians agree that the first major event to herald the Great Awakening took place in Northampton in November and December of 1734 during a series of sermons of Jonathan Edwards on justification by faith alone. According to Edwards's personal records, we know "the first indications were visible in the fall of 1733, when the young people began to show an unusual flexibleness and yielding to advice."[6] The first spectacular conversion of major consequence, however, took place a year later in December 1734, when a "notorious" woman surprised both Edwards and Northampton by presenting unmistakable signs of election that clearly qualified her for full church membership.[7] The news of this woman's conversion served as a catalyst for many other experiences of God's grace in Northampton, especially among its young people.[8]

Events did not stop then or there. The revival grew in Northampton, rippling outward from town to town along the Connecticut River valley through the spring and summer of 1735.[9] "By March and April of 1735," as George Marsden describes these events, "the spiritual rains had turned the stream to a flood."[10] In Northampton alone, church membership nearly doubled, growing by three hundred in three months to include most adults in the town.[11] Edwards continued to be both surprised and pleased by these events, which he carefully observed and recorded. The account he published of that early revival, *A Faithful Narrative of the Surprising Work of God in the Conversion of Many Hundred Souls in Northampton, and Neighboring Towns and Villages* (1737), especially his skillful descriptions of the types and stages of the conversion experience, had a major impact on revival expectations in Europe and America, alerting "both Old England and New to the possibility and process of remarkable conversion."[12] John Wesley read *A Faithful Narrative* that same year, remarking, "Surely this is the Lord's doing, and it is marvelous in our eyes."[13] A year later in 1738, both Charles and John Wesley went through individual conversion experiences, after which they, along with George Whitefield, initiated the influential Methodist revivals in England.[14] Edwards's account also influenced George Whitefield, who considered his own enormous English crowds to be inspired by the same Spirit that had fanned the spark of life into the "great awakening" of Northampton.[15]

Although the Northampton regional revivals of 1734–1735 together amounted to one of the largest events of church renewal so far in New England's history, they were but a preview of the enormous events of the major revival period of the Great Awakening. This major revival period is usually considered to have started in the fall of 1740 with the arrival in New England of George Whitefield, the charismatic English Methodist evangelist minister and first of the Great Itinerant preachers. Whitefield first preached in the fall of 1739 to large audiences in Philadelphia, moving later that year to Savannah, Georgia, then working his way north to arrive in Boston in mid-September 1740, at times preaching to crowds as large as twenty thousand.[16] With his arrival in New England, Whitefield carried the Northampton revival spark full circle. As the fervor spread North through the colonies, as well as "through every element of the populace," the revivals and conversions he left in his wake were together hailed as a "great and general awakening," which would form the standard by which all the many larger waves of revivals that would pulse through America in the next centuries were measured.[17] Whitefield was soon followed by other itinerant preachers, particularly Gilbert Tennent, a Presbyterian minister from New Jersey who also drew enormous crowds.

The phenomena of the major revival period of the Great Awakening differed in several important respects from the many isolated and scattered, local and regional revivals of the previous one hundred years. The first major difference was the unprecedented scope and scale of these major revivals. Previous revivals were usually circumscribed by a particular town and its surrounding communities, often limited naturally by geographical access, as was the case in Edwards's 1734–1735 revival, which was largely confined to the Connecticut River valley. In contrast, the major revival period of the Great Awakening involved most of the American colonies, especially the Middle and New England colonies.[18] In New England alone one hundred and fifty towns experienced large revivals.[19] The vast size of the crowds—often tens of thousands—who thronged to hear the Great Itinerants speak had never been seen before.[20] Both Boston and Philadelphia recorded events attended by over twenty thousand—a staggering number considering that in 1740 Boston was probably the largest colonial city with approximately seventeen thousand residents.[21] Another measure of the size of the Great Awakening was the increase in membership that New England churches saw in its wake. The Sunday before Jonathan Edwards gave his famous *Sinners in the Hands of an Angry God* sermon at Enfield, the neighboring Suffield congregation experienced a revival in which church membership expanded by ninety-five souls on that day alone.[22] In Boston during late 1740 and early 1741, the effects of Whitefield's and Tennent's preaching was dramatic, with "many hundreds of souls" during this period joining the company of the elect.[23] In New England overall, even with more stringent requirements to provide proof of a true conversion experience, church membership has been estimated to have grown by as much as fifty thousand.[24]

The second major difference was the highly emotional tenor of the revivals themselves, where "flamboyant and highly emotional preaching made its first widespread appearance in the Puritan Churches."[25] Even the nature of the conversion experience was different: decidedly more intense, highly emotional, and at times quite physically dramatic. Ahlstrom reasons that to comprehend this aspect of the phenomenon, especially in Puritan New England, we must remember that church worship, including the call to revival, had grown to a "staid and routine formalism in which experiential faith had been a reality to only a scattered few."

The peak of the Awakening's revivals and its waves of conversions took place during the years 1740–42, with "an aftermath of violent controversy" appearing immediately in the wake of Whitefield's first three-month tour of New England in the autumn of 1740.[26] That controversy was focused generally on Whitefield's preaching style and the highly emotional revivals and conversions his sermons produced. And as the revivals continued to grow and spread, that controversy continued to mount through 1741 and 1742, so that 1743 was spent in "highly polemical evaluations."[27] It was, of course, New England that not only experienced the greatest effects of the Great Awakening but also waged the fiercest debate concerning it.[28]

There were of course many facets to the controversy that emerged, but the central concern of the Awakening's major critics was whether these revivals, the preaching that catalyzed them, and the many conversion experiences that resulted from them were in fact genuine works of the Spirit. The intense theological criticism that came was led by Charles Chauncy of Boston, the anti-revivalist champion of a conservative orthodoxy of "reasonableness" known as Old Lights.[29] The main criticism of this anti-revivalist camp was that "far from being a supernatural work," the conversions were "criminally excited by artificial stimultions ... under a pretense that God himself was working the harm."[30] Edwards was among the most prominent leaders of the New Lights, the pro-revivalist faction, who maintained that the revivals and their conversions were clearly a manifestation of God's will at work in the world. But as the excesses of the Awakening became more apparent, even Edwards became increasingly critical of certain revival attitudes and practices, condemning clear cases of excess and even personally rebuking Whitefield at one point. Throughout the controversy, however, Edwards maintained that the revivals constituted a genuine work of God, which should be fostered even as it should also be purified. While the Awakening was at its peak, he both defended and criticized the revival in *The Distinguishing Marks of a Work of the Spirit of God* (1741) and *Some Thoughts Concerning the Present Revival of Religion in New England* (1742). After the revival frenzy had died down, he published perhaps his most influential work, *A Treatise Concerning Religious Affections* (1746).

In these works, Edwards argued against the ideal of sober, "reasonable" religion, presenting the theme that most distinguishes his theology, that the essence of all true religion lies in holy love, a love that proves its genuineness in a truly regenerated person by its internal quality and its external, practical results. By emphasizing holy love in his theology of conversion, Edwards addressed one of the primary intellectual issues of the debate between New Lights and Old Lights concerning human psychology: Is true religious experience rational, based on reason, which should ideally govern the will and the emotions? Or is the experience of God's grace primarily a matter of the heart, the will, and at its core emotional? While it is certainly true that the debate between religion of the head and religion of the heart has a extremely long heritage within Christianity, it is also true that within the context of the Puritan movement and the American colonies during the Great Awakening these issues came to the fore in such a way as to alter the larger conceptual framework and cultural world view. To understand the nature of this controversy and the profound conceptual shift that occurred during these debates, we need to explore the thought of Jonathan Edwards in greater detail.

The Conceptual Revision of Jonathan Edwards

Jonathan Edwards was born in 1703, one year before the death of John Locke and the publication of Newton's *Opticks*.[31] In 1716, Edwards entered Yale College where he encountered—likely through his tutor, Samuel Johnson—John Locke's *Essay Concerning Human Understanding* (1691) and Isaac Newton's *Principia Mathematica Philosophiae Naturalis* (1687) and *Opticks* (1704). By 1723 at age twenty, Edwards had completed two notebooks, "The Mind" and "Notes on Natural Science," in which he had outlined a system of philosophy that incorporated a great many of the new thoughts of these two great thinkers.[32] The ideas worked out in these notebooks formed a highly developed theological and philosophical system that informed all his later works. While the core of Edwards's philosophical synthesis in these two notebooks was completed relatively early in his life, they were not published while he was alive. His influence on Puritan theology, therefore, came through his later writings. Although Edwards was first prominent in New England circles as the heir to his grandfather, Solomon Stoddard, it was in the events and controversies that arose from his own Northampton revival of 1734–1735 and even more from the Great Awakening's major revivals of 1740–1742 that his new formulations would become most influential and his reputation as the leading theologian in New England, indeed in the American colonies, would be secured.[33]

Whereas the encounter with the ideas of Locke and Newton led his tutor, Samuel Johnson, to abandon Puritanism for the more liberal, Anglican ministry, their ideas led Edwards in a quite different direction.[34] Recognizing in these new ideas a grave potential danger not only to Puritanism but to religion generally, Edwards devoted himself to revising Puritan philosophy and theology so that it might endure by demonstrating that "far from negating the old doctrine" these new ideas "actually supported it."[35]

Edwards embraced Newton's system of ideas completely, including his heliocentric cosmology and such seminal concepts as infinite space as God's sensorium, in which God's infinity and omnipresence constitute and maintain both duration and space. As part of God, duration and space are the means by which God perceives and comprehends all things that exist in time and space.[36] Equally important was Edwards's enthusiastic adoption of Newton's atomistic and mechanistic model of physical reality, since that system's mechanical determinism resonated so strongly and could be used to support so squarely his own life-long emphasis on the doctrine of predestination. And although Edwards recognized the dangerous implications of God's diminished role inherent in the clock-work universe so easily derived from Newton's thought, he steadfastly rejected the common Deist interpretation, which held God to be the creator and architect of a mechanical universe that no longer required divine involvement or intervention.

For our purposes, Edwards's response to Locke's empirical psychology was more involved and complex than his integration of the ideas and consequences he found in Newton. That Locke's epistemology and psychology presented a completely different model than the Puritan scholastic system of multiple, hierarchical faculties depending on phantasms and supernatural mediating agents described earlier was as obvious to him as were its alarmingly disenchanting implications. As Newtonian physics opened wide the door for eliminating God's continuous role in maintaining the existence and rationality of the universe, so Locke's empirical epistemology eliminated the need for God's agents from the process by which we apprehend that universe. Edwards's primary intent was to integrate both of these new systems of thought into Puritan theology in such a way that he could maintain the essential Puritan doctrinal triumvirate while he simultaneously reaffirmed God's continual and personal presence in the world. In this integration, Edwards arrived at a position of immaterial, or subjective, idealism that was similar in many important respects to the philosophical system constructed by Bishop George Berkeley a decade earlier in response to these same ideas.[37]

Edwards found the way to this position through what he saw as a logical consequence of combining Newtonian atomism with Locke's distinction between simple and complex ideas.[38] Early in *An Essay Concerning Human Understanding,* Locke denies the claims made by many philosophers such as Plato and Descartes that the human mind is naturally endowed at birth with certain innate ideas. Instead, the human mind at birth is a tabula rasa, on which sensory experience alone writes knowledge. The ideas of all human minds derive from two processes: sensation and reflection. Sensation is passive and provides specific, individual ideas about particular sensory objects, such as the whiteness and spherical shape of a billiard cue ball. Reflection is an active ability of the mind to observe our own mental processes and to combine the ideas derived from sensation. Locke also described two broad categories of ideas: simple ideas derived directly from sensory experience of objects, and complex ideas. By reflection on the simple ideas of sensation, the human mind is able to combine those simple ideas to form more complex ideas, such as generalizations, in much the same way that Newtonian atoms combine to form larger, more complex substances. Locke generally seems to hold an absolute distinction between ourselves as perceivers of external objects and the perceived objects themselves; and he is usually interpreted as describing a representational theory of perception and knowledge, in which the mind does not directly perceive external objects. Instead, the mind directly perceives the simple ideas of sensation, which act as representations of particular external objects.

Edwards starts his synthesis with Newtonian concepts about solid bodies and an atom as a physical minimum (that which cannot be divided further), first defining body as solidity, and solidity as the power to resist another body occupying the same space. He then asks, "If body is solidity, and solidity is simply the power of resistance, what is it that resists?"[39] Most importantly, what could such resistance mean and how could it operate in the absence of mind? Take "two globes only existing; and no mind," Edwards begins.

> There is nothing there, *ex confesso,* but Resistance. That is, there is such a Law, that the space within the limits of a globular figure shall resist. Therefore, there is nothing there but a power, or an establishment. And if there be any Resistance really out of the mind, one power and establishment must resist another establishment and law of Resistance, which is exceedingly ridiculous.[40]

In one sense, Edwards is maintaining that it makes no sense to say that one power of resistance (the first globe) could resist another power of resistance (the second globe), since that would imply that resistance resists resistance, which is meaningless. In another sense, though, Edwards does not find such thinking logical because he still resides partially in the medieval-scholastic world view, which holds that all motion occurs by means of the will or mind of some conscious spiritual agent. For Edwards, because resistance is a power or motive force, it can exist only in the mind.

At this point, Edwards finds help in Locke's concept of mixed modes.[41] Mixed modes are complex ideas formed by combining simple ideas of different kinds, as in the case of the idea of beauty, which is formed by combining simple ideas of color, form, texture, and other features that in their combination produce pleasure in the observer. The idea of beauty, however, does not exist anywhere but in the mind of the observer. Unlike simple and complex ideas, such as the idea of a white ball, the idea of beauty is not a representation of a specific substance or object that exists in the external world. Rather, beauty is the idea of a mode, or manner of expression, that is experienced through a great number of substances. The idea of beauty, therefore, exists solely in the mind.[42] Taking up this concept of modes and their existence solely in the mind, Edwards continues.

> But now it is easy to conceive of Resistance, as a mode of an idea. It is easy to conceive of such a power, or constant manner of stopping or resisting a colour. The idea may be resisted, it may move, and stop and rebound; but how a mere power, which is nothing real, can move and stop, is inconceivable, and it is impossible to say a word about it without contradiction.[43]

But Edwards previously defined body as solidity, with solidity meaning nothing more or less than this power of resistance. Resistance, though, is a mode of an idea, making body (resistance) an idea existing only in the mind. "The world is therefore an ideal one," Edwards concludes, "and the Law of creating, and the succession, of these ideas is constant and regular."[44]

As ideas, the existence of the world and all the objects in it are utterly dependent upon mind's continuous perception of them. And because the continuous perception of any one thing (let alone all things) could not come about by human agency, this conclusion opened wide the gate for both spirits and, most especially, God to pervade the universe once again. For while some of the universe might be maintained by the perception of particular spirits and human beings, clearly the "whole of it can be maintained by no other agent but God," so that

> this substance of bodies at last becomes either "nothing" or nothing but the Deity acting in that particular manner in those parts of space where He thinks fit. So that speaking most strictly there is no proper substance but God Himself.[45]

Edwards's immaterialism, like Berkeley's, concludes that God alone is real substance. Because God is everywhere and at all times fully immanent and omnipresent, however, physical reality is guaranteed by the will and mind of God. In this model, Newtonian mechanistic determinism, Calvinistic predestination, and the unified psychology of Locke are all one thing—the mind of God. And thus we are back at the essential Puritan position, the absolute sovereignty of God, recast in modern philosophical and psychological terms. To his own judgment, Edwards succeeded.

Edwards's revision of Puritan theology was complex and comprehensive. It is not possible in this study to present even a survey of all that he accomplished. Throughout his career, he was interested in preserving what he regarded as the three foundational doctrines of not just Puritanism but of all true religion, the three we have seen throughout this study: God's absolute sovereignty, human depravity, and predestination. Edwards's synthesis, especially his immaterialism, allowed him to change the emphasis of many aspects of these three doctrines, reformulating them along lines that would keep them alive through the eighteenth and early nineteenth centuries in New England Calvinism's long battle between orthodox Puritan theology and Edwards's revised theology on the one hand and, on the other, its efforts to stem the advancing tide of more liberal theologies, especially those of Arminianism, Deism, and Unitarianism.

The theory that Edwards developed on the nature of the Trinity may serve as a particularly striking example of his reformulation—one that enabled him to conceive anew the nature of the conversion experience along the lines of what he observed during his own revivals and the Great Awakening.[46] Developed for the first time fully in his early "Notes on Natural Science," his theory of the Trinity was based on several concepts taken from Locke's epistemology, including the concept of introspection, or reflection, as a source of knowledge, and the important distinctions between the perceiving subject and perceived object, and between the thing perceived and the representational idea of the thing perceived. With these Lockean concepts in mind, Edwards starts, however, quite characteristically with the basic Puritan theological doctrine of the omniscience of God: since God knows all there is and God's knowledge is perfect, God must know himself perfectly. According to Edwards's idealism, a perfect idea in the mind of God is the thing itself, so that through self-reflection, this perfect idea that God has of himself is a "substantial" image, which is at one with his own essence and properly God. Through self-reflection, therefore, God has begotten another being who is a perfect and substantial image of himself. In this way, the Son is born from the Father. And because God is all excellency, virtue, and perfect self-love, his love at seeing his own image becomes the Spirit. The Godhead is therefore divisible into two Lockean entities: God as perceiving subject (God the Father and Creator), and God as perceived object (God the only begotten Son, the image and word of the Father). The relationship between the two is the Spirit, or God's love of God himself. This affectional concept of the Holy Spirit enabled Edwards to recast the process of regeneration by the Spirit in a more joyful, more emotional, affectional, love-based, and heart-centered vein than was common in many previous, reason-centered Puritan theologies.

For New England divines who considered religious experience to be a process of reestablishing the proper hierarchy of psychological faculties, the enthusiasm and emotionalism of the Great Awakening revivals and conversions made them suspect. Charles Chauncy in numerous publications like *Enthusiasm Described and Caution'd Against* (1742) and *Seasonable Thoughts on the State of Religion in New England* (1743) made clear the orthodox, Old Light, position:

> There is the Religion of Understanding and Judgment, and [that of the] Will, as well as of the Affections; and if little Account is made of the former, while great Stress is laid upon the latter, it can't be but People should run into Disorders.[47]

Chauncy's assessment is based on the dominant scholastic faculty psychology described earlier, in which the superior faculty of understanding would ideally govern the will, which would in turn control the affections. To these Old Light theologians, things were as they should be if the will were properly under the control of the understanding. In contrast, "any appeal directly to the passions, which attempts to bypass [the understanding], was demonstrably immoral," since uncontrolled emotions were a certain sign of an individual's fallen state continuing to operate.[48]

Edwards responded to these criticisms on several levels in *Some Thoughts Concerning the Present Revival of Religion in New England*. First, to those for whom "the affections of the soul are something diverse from the will, and not appertaining to the noblest part of the soul, but the meanest principles that it has," Edwards provided a resounding affirmation of the central role of religious affections:

> I cannot but think that these gentlemen labor under great mistakes, both in their philosophy and divinity. It is true, distinction must be made in affections or passions. . . . But it is false philosophy to suppose this to be the case with all exercises of affection in the soul, or with all great and high affections; and false divinity to suppose that religious affections do not appertain to the substance and essence of Christianity: on the contrary, it seems to me that the very life and soul of all true religion consists in them.[49]

The emotional nature of the revival conversions, therefore, was completely appropriate and true to genuine religious experience. Not only did Edwards define and defend this religion of the heart as the New Light response to Old Light orthodoxy, but we can also see in this statement and those that follow how deeply the encounter with Locke's epistemology had convinced Edwards that the then-dominant psychological model of separate, hierarchical faculties was inadequate to describe human nature.

To the criticism that the Awakening was at fault for not directing efforts to the understanding, Edwards responded with a direct critique of the Old Light psychology, offering instead a more unified, organic alternative:

> I humbly conceive that the affections of the soul are not properly distinguished from the will, as though they were two faculties in the soul. All acts of the affections of the soul are in some sense acts of the will, and all acts of the will are acts of the affections. All exercises of the will are in some degree or other, exercises of the soul's appetition or aversion; or which is the same thing, of its

love or hatred. The soul wills one thing rather than another, or chooses one thing rather than another, no otherwise than as it loves one thing more than another; but love and hatred are affections of the soul: and therefore all acts of the will are truly acts of the affections.[50]

Edwards's model here is that of a unified human soul in which "understanding and cognition occur to the entire being as it is 'affected,' and that includes the sense of the heart in which the highest kind of knowledge, spiritual knowledge, is centered."[51] Understanding, will, and affection to Edwards are organically connected, simultaneously acting, and inseparable except by meaningless intellectual categories. True religion, he reaffirms, is of the heart:

> All will allow that true virtue or holiness has its seat chiefly in the heart, rather than in the head: it therefore follows, from what has been said already, that it consists chiefly in holy affections. The things of religion take place in men's hearts, no further than they are *affected* with them.[52]

Yet another piece of the transition that Edwards's psychological model represents is visible in his clear move away from the scholastic agents of transmission, or phantasms, to Locke's empirical representational theory of ideas in his defense of passionate preaching as a more effective way of representing and therefore conveying religious matters:

> I think an exceeding affectionate way of preaching about the great things of religion, has in itself no tendency to beget false apprehensions of them; but on the contrary a much greater tendency to beget true apprehensions of them, than a moderate, dull, indifferent way of speaking of them. An appearance of affection and earnestness, in the manner of delivery . . . has so much the greater tendency to beget true ideas or apprehensions in the minds of hearers, of the subject spoken of, and so to enlighten the understanding . . . and that for this reason . . . does in fact, more truly represent them. . . . If a subject be in its own nature, worthy of very great affection, then a speaking of it with very great affection, is most agreeable to the nature of that

Beyond a Disenchanted Cosmology

subject, or is the truest representation of it, and therefore has most of a tendency to beget true ideas of it, in the minds of those, to whom the representation is made.[53]

Because ideas in the mind represent the objects of our understanding, emotional matters of the heart must be conveyed accurately by using the appropriate emotions. Otherwise, matters of the heart cannot truly be understood. And because the sovereign God's will is nothing if not pure and holy Love, matters of religion and true religious experience must therefore be a matter primarily of the heart and in their essence emotional.

One of the transitions evident in the thought of Jonathan Edwards, as he engaged this theological controversy while simultaneously integrating his observations of the Great Awakening, concerns the psychological theory of human nature and human behavior. Edwards's thought has moved significantly away from the medieval conception of man composed of a hierarchy of damaged faculties, dependent upon other agencies for knowledge, to a more modern one involving coherent responses and choices, in which all capabilities are seated in an organically unified human being, and for which the best model is love.

As Ahlstrom, for example, concludes,

> The "affections" to Edwards are not simply the emotions, passions, or even the "will," but more fundamentally, that which moves a person from neutrality or mere assent and inclines his heart to possess or reject something.[54]

For Edwards, love "is not only one of the affections but it is the first and chief . . . and the fountain of all affections."[55] And as Edwards conceived it in his theory of the trinity noted earlier, love in its purest form is the will of the sovereign God.

The revision Edwards accomplished was part of a much larger conceptual transformation in America that involved major changes in psychology, philosophy, and theology. Edwards played a major role in this cultural transformation and larger paradigm shift. It was not so much that he alone changed the Puritan world view but that he was one of the primary voices for a new system of thought that looked at the world, at religion, and at the human being in a radically different way. Implicit in all of his arguments during the Great Awakening was a new psychological and epistemological viewpoint that broke from a medieval conception of human beings as composed of a hierarchy of damaged psychological faculties, that demoted reason from its honored placed of authority, and that emphasized the emotional response and its integration with reason and the understanding. We can see in this debate a foreshadowing of the one that will arise in the late eighteenth and early nineteenth centuries between the Enlightenment emphasis on rationality and the Romantic honoring of the imagination, emotionalism, and the transformation of the human individual through struggle, creativity, and an immersion in Nature. We also find in Edwards's thought one of the more important foundational statements of a psychological approach to the study of human nature and the religious conversion experience the thread of which weaves through at least the work of William James, which in turn has had such enormous influence in later transpersonal psychological models.[56]

According to nearly every authority, it would be difficult to overestimate the importance of the Great Awakening.[57] In significant ways, the Great Awakening and Edwards's theological revision broke the hegemony of the monolithic Puritan orthodoxy that had held power in New England and much of the Middle colonies for over a century. One of the major, immediate effects of the Great Awakening, quite apart from Jonathan Edwards's theological revision, was its rapid democratization of the conversion process, making it available to anyone from any walk of life. The events of the many huge revivals of the Great Awakening changed the rules of conversion. Suddenly spiritual awakenings were happening everywhere to anyone, Half-Way covenanted or not. Arduous, disciplined preparation was not necessary; studious bible study made little difference. After white hot torment, conversion came in a sweet flood of psychological and spiritual release regardless of individual biographical details. Regeneration required only the catalytic words of a godly minister—a George Whitefield, a Gilbert Tennent, or a Jonathan Edwards—able to spark the purifying conflagration that came through the presence of the Spirit. But it was this very shift in the nature and availability of the conversion process that made it a fruitful topic for Edwards's revision, thereby rendering the religious conversion process something that modern depth psychology movements could examine in great detail and incorporate into future practices.

While Edwards's theological revisions were not directly responsible for the democratization of the conversion experience, his statements regarding the nature of the religious conversion experience helped ensure that it would be thought of as available to anyone independent of previous church affiliations. Moreover, in the overall transition that was made during and after the Great Awakening, Edwards's theology played an enormously important catalytic role. At first, his position was the liberal, revolutionary reaction, the New Lights, to the conservative Puritan orthodoxy, the Old Lights, as represented by Charles Chauncy. Edwards's successors would be known as the New Divinity, and from their continuing battle with Puritan orthodoxy another more liberal position would very quickly evolve, shifting Edwards's successors to the conservative side. This third theological faction would be "a liberal rationalist theology," which was more in line with the rationalism of the Enlightenment and with what would emerge as "the Liberal theology of the rest of the 18th century."[58]

The controversies of the Great Awakening would continue to be debated into the early nineteenth century. The divisiveness of Old Lights and New Lights would eventually drive many people away from the Puritan conservatism of both Lights into the more liberal churches, contributing significantly in New England to the success of the Arminian tradition in a revivified Episcopal Church, the emergence of Unitarianism as the dominant church for much of the nineteenth century, the development of a religion of nature, and the Transcendental movement centered around Emerson.[59] Moreover, these intense controversies were one of the primary means by which the Great Awakening became the "psychological earthquake in the human landscape" that propelled New England and America out of a medieval intellectual framework into one more closely aligned with the modern era and Enlightenment developments in England and continental Europe.[60]

Whereas the list of repercussions and consequences could go on for many pages, as it does in many sources, one of the more intriguing aspects of the Great Awakening is that no consensus seems to have been reached on why it took place when it did.[61] On the one hand, many historians offer theories emphasizing the causative role of particular social or economic forces. Alan Heimert and Perry Miller, for example, present a fairly confident theory based on economic and social class distinctions:

What was at work throughout the Western world is fairly obvious: the upper or the educated classes were tired of the religious squabbling of the seventeenth century, and turned to the more pleasing and not at all contentious generalities of eighteenth-century rationalism; the spiritual hungers of the lower classes or of what, for shorthand purposes, we may call "ordinary" folk were not satisfied by Newtonian demonstrations that design in the universe proved the existence of God. Their aspirations finally found vent in the revivals.[62]

Such approaches often provide contradictory explanations, though, rendering the enterprise unsatisfactory. Ahlstrom, for example, finds no agreement among historians and no single theory capable of explaining the phenomenon:

Given the immense social consequences of the awakening, it is unfortunate that historians have been almost totally unable to agree as to why it occurred. Few if any improvements have been made on the conflicting explanations advanced by Edwards and Chauncy in the mid-eighteenth century. It came to pass that the message of sin and redemption spoke with sudden and peculiar power to the condition of many New Englanders. . . . Efforts to establish political, social, and economic explanations have been conflicting and unconvincing. Revivals and conversions seem to have come to churches and persons in all areas and classes and walks of life. Given present limitations of knowledge, it is enough to say that in New England as elsewhere the revivals became a major means by which people of many diverse types responded to changing moral, religious, intellectual, and social conditions.[63]

From an archetypal perspective, however, the events of the Great Awakening took place during the same years that Uranus and Neptune formed an opposition alignment (1728–1746). The entire period presents a wealth of phenomena with characteristics that are remarkably consistent with central themes of the Uranus-Neptune complex. And beyond these Uranus-Neptune correlations, specific events within the Great Awakening period present a notable pattern of correlations with the timing of Jupiter and Saturn alignments with the three outermost planets that are also consistent with the archetypal themes associated with these alignments.

An Archetypal Perspective on the Great Awakening

Figure 1 illustrates the configuration of the five outer planets in December 1734. Figure 2 illustrates the configuration six years later, in October 1740, near the beginning of the major revival period of the Great Awakening. Table 1 presents the start, exact, and end dates for various dynamic alignments of the outer planets that formed during the period of the Uranus-Neptune opposition that began in 1728 and ended in 1746. The two figures clearly illustrate the opposition alignment of Uranus and Neptune that dominated the entire period of the Great Awakening.

Figure 1 **Outer Planetary Configuration December 1734**

1734 December
Northampton, Massachusetts

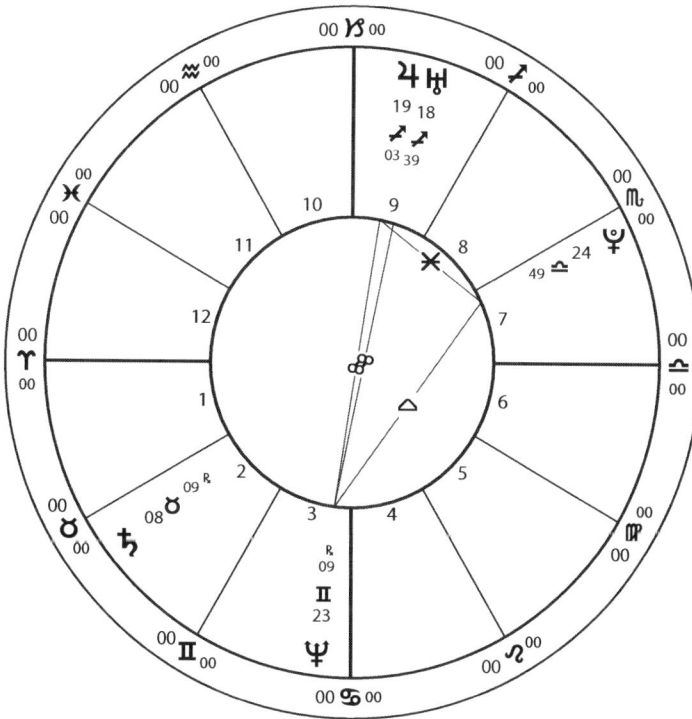

Figure 2 **Outer Planetary Configuration October 1740**

1740 October
Boston, Massachusetts

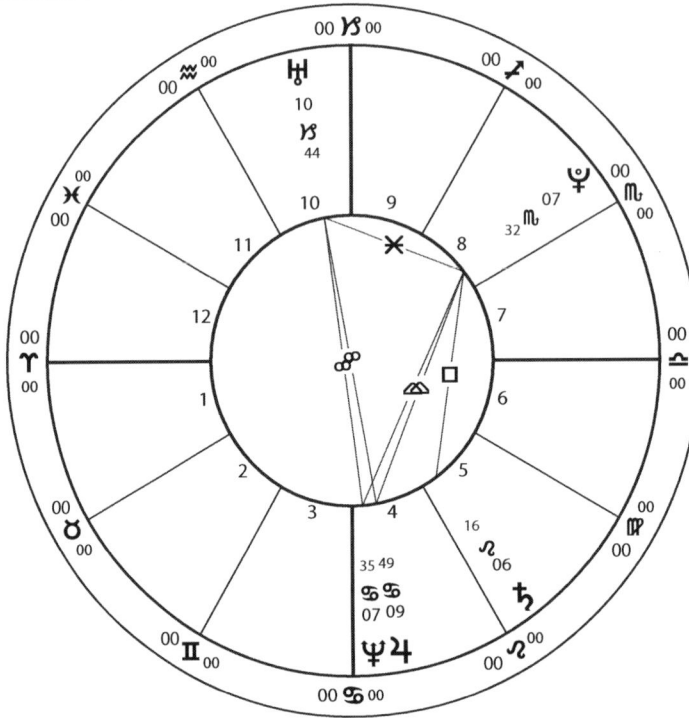

Table 1   **Outer Planetary Alignment Dates (1728–1746)**

Alignment	Start	Exact	End
Uranus opposite Neptune	1728	1734–1738	1746
Jupiter conjunct Uranus	Jan 1734	Nov–Dec 1734	Oct 1735
Jupiter opposite Neptune	Jan 1734	Dec 1734	Nov 1735
Jupiter conjunct Neptune	Jun 1740	Sep 1740–Apr 1741	Jul 1741
Jupiter opposite Uranus	Jul 1740	Oct 1740–Jun 1741	May 1742
Saturn square Pluto	Jul 1740	Oct 1740–Jul 1741	Jul 1742

As shown in figure 1, the Uranus-Neptune opposition was within 5° of an exact alignment in December 1734, when the first spectacular conversion took place in Northampton. The Uranus-Neptune opposition was exactly aligned several times during a five-year period beginning in early 1734 and ending in 1738. During this same five-year period, the Northampton revival began (1734), Jonathan Edwards published his account of that revival *(A Faithful Narrative,* 1737), George Whitefield began his sensational preaching career in England (1737), John and Charles Wesley underwent individual conversion experiences (1738), and two globally significant new religious movements, Methodism and modern Evangelicalism, were born (1738).[64]

As shown in figure 2, at the height of George Whitefield's first New England tour in October 1740, Uranus and Neptune were approximately 3° away from an exact opposition alignment. By 1745, the spectacular revivals and mass conversions of the Great Awakening in America had largely subsided in synchrony with the waning archetypal intensity of the Uranus-Neptune complex as the two planets moved beyond an active orb (15°) for the first time in early 1744 and for the final time in late 1746.[65]

As in many other historical periods during which Uranus and Neptune have formed major hard aspects, the 1730s and 1740s in America and Britain were marked by a heightened awareness of spiritual and religious matters, a revitalization of religion and spirituality in general, and reports of large numbers of individual experiences of spiritual awakening, religious conversion, and altered states of consciousness.[66] Such phenomena display typical themes of the Uranus-Neptune complex, combining Neptune's associations with the spiritual, mystical, transcendent, and imaginal realms with Uranus' associations with sudden breakthrough and lightning-like illumination that crack open established structures, shatter barriers, and explode previous definitions to yield sudden brilliant insights, profound shifts in understanding, spiritual awakening, and religious conversion. As Edwards described the early events of the 1734 revival, the first spectacular conversion was "almost like a flash of lightening, upon the hearts of young people all over the town, and upon many others."[67]

Beyond these central Uranus-Neptune correlations, as we have seen in the events surrounding the birth of Puritanism and the Half-Way Covenant crisis, a closer examination of the timing and characteristics of specific events and controversies within the Great Awakening period presents a notable pattern of correlations that are consistent with other dynamic planetary alignments of this period. The timing of Jupiter alignments with this Uranus-Neptune opposition correlates with specific months of the early revival activity (see table 1). Jupiter began to conjoing Uranus and oppose Neptune in early 1734, the same period that the Uranus-Neptune opposition reached an exact alignment for the first time. This Jupiter-Uranus conjunction was exactly aligned for the first time during November and December of 1734, when Jonathan Edwards's sermons began to stir the crowds. In December 1734, Jupiter was both exactly conjunct Uranus and exactly opposite Neptune. All three alignments, therefore, were exact or nearly exact during December 1734, when the first spectacular conversion experience set Northampton and the Connecticut River valley ablaze with revivals. Both the Jupiter-Uranus conjunction and the Jupiter-Neptune opposition ended in the autumn of 1735, by which time that region's revival fervor had subsided. In combination with other archetypes, the Jupiter archetype typically brings associations with expansion, elevation, and success. In this case, Jupiter was aligned with the Uranus-Neptune opposition with its archetypal associations with altered-consciousness, religious awakening, and spiritual conversions. The Jupiter-Uranus archetypal combination in particular is associated with successful breakthroughs and surprising turns of events in positive directions.

Five years and one half-cycle later, Jupiter again moved into axial alignment with this Uranus-Neptune opposition, this time conjunct Neptune and opposite Uranus. These Jupiter alignments started just two months before the major revival period of the Great Awakening began: the Jupiter-Neptune conjunction started in late June 1740 and was exact in early September; the Jupiter-Uranus opposition started in July 1740 and was exact in October and November of 1740. Whitefield arrived in New England in September 1740, when he preached to enormous crowds starting in Boston, sparking the full Awakening in New England. In October of 1740, Whitefield arrived in Northampton to visit Jonathan Edwards and preach to his congregation and begin the second wave of revivals there.[68] Both Jupiter alignments were exact again in the spring of 1741, with the Jupiter-Uranus conjunction ending in May 1742. The months from the fall of 1740 through the summer of 1741 marked a time in the Great Awakening that was full of revival enthusiasm with large, exultant crowds, accurately matching the period during which Jupiter was aligned with the Uranus-Neptune opposition.

Important differences between these two major revival periods are also apparent. Even as the enthusiasm and optimistic momentum of the Awakening continued to build from late 1740 well into 1741, a strong pattern of controversy simultaneously emerged during this period that was not present in the Northampton revivals of 1734 through 1735. In contrast to the earlier revival period, the events of 1740–1742 present numerous phenomena consistent with the timing and archetypal themes of the Saturn-Pluto square alignment that also formed during the Great Awakening. This Saturn-Pluto square alignment started in July 1740, was exact during the period from October 1740 through July 1741, and ended in July 1742 (see table 1), which is precisely the period during which the revival events became disruptive, conversion experiences grew more emotional and dramatic, intense criticism emerged, and the period's central controversies were fully engaged.

In other words, as Jupiter moved into alignment with the Uranus-Neptune opposition during both periods, characteristics typically associated with the Jupiter archetype such as liberation, enthusiasm, optimism, joy, success, and expansion appeared in combination with the spiritual-awakening, religious conversions, and shifts in consciousness typically associated with periods of major Uranus-Neptune alignments. Simultaneously, the whole complex set of phenomena became heavier, more profound, more intense and passionate, more problematic and controversial, displaying characteristics typically associated with Saturn-Pluto periods, such as obsessive emotional intensity, hellfire and brimstone preaching techniques, intense moral scrutiny and criticism, and the uncontrollably dramatic birth and death contractions of regeneration that are evident in the "intensity of bodily effects of conversion—fainting, weeping, shrieking."[69] Such phenomena began to appear and were prominent features of the Great Awakening during the 1740–1742 period in synchrony with Saturn and Pluto as they moved into square alignment starting in July of 1740, were exactly aligned starting October of 1740, and moved out of orb in July 1742.

As with the highly controversial nature of other historical periods examined so far that took place while dynamic Saturn-Pluto alignments have formed, the early events of the major revival period of the Great Awakening rendered it controversial almost immediately, resulting in intense debate between highly polarized factions. The controversies that emerged focused on two major issues: first, concerning accusations made by Whitefield, Tennent, and other itinerant preachers that many of New England's established ministers were unconverted, placing both individuals and entire congregations in extreme danger; and second, concerning the unprecedented emotionalism and enthusiasm of the individual revival events that comprised the larger awakening, including the highly dramatic sermons of this period as well as the disruptive and histrionic nature of many revival events and individual conversion experiences.

Soon after he arrived in New England, Whitefield generated one of the major controversies of the Great Awakening by negative comments he made during his New England tour in late 1740 (published in April 1741), concerning the unconverted status of many of New England's ministers.[70] The controversy intensified on October 21, 1740 as a result of Whitefield's sermon at Suffield, Massachusetts, in which he emphasized the danger congregations faced by supporting and listening to ministers who had not themselves experienced regeneration.[71] Following Whitefield's departure from New England, Gilbert Tennent continued to emphasize this theme during his own hugely successful revivals in Boston from late 1740 through early 1741, especially as formulated in his famous sermon *The Danger of an Unconverted Ministry,* first delivered in 1740 then published in both Philadelphia and Boston in 1742.[72] In this sermon, Tennent calls anti-revivalist ministers established in congregations the enemies of Christ, who like the

> Pharisee-Teachers, having no Exprience of a special Work of the Holy Ghost, upon their souls, are therefore neither inclined to, nor fitted for, Discoursing, frequently, clearly, and pathetically, upon such important Subjects. . . . All the Doings of unconverted Men, not proceeding from the Principles of Faith, Love, and a new Nature, nor being directed to the divine Glory as their highest End, but flowing from, and tending to Self, as their Principle and End; are doubtless damnably Wicked in their Manner of Performance, and do deserve the Wrath and Curse of a Sin-avenging God.[73]

Tennent and other itinerant preachers who took up this theme encouraged congregations to remove established ministers with anti-revivalist sentiments. If removal was not possible, individuals should leave their congregation for one led by a pro-revivalist.

This tendency to judge systems of belief according to black and white categories, in which one position is absolutely good and the opposite is immoral, sinful, or evil, is characteristic of dynamic Saturn-Pluto alignment periods. Jonathan Edwards considered the intense censorial judgment of this issue to be the most tragically misguided of the many consequences of the Great Awakening. In 1742 with Saturn and Pluto still in square alignment, Edwards specifically drew attention to the inimical effects of this tendency to project evil, sin, and wrongdoing onto others:

> The 1ˢ Thing I would take notice of, is, censuring others that are professing Christians, in good standing in the visible church, as unconverted. I need not repeat what I have elsewhere said to show this to be against the plain . . . and strict prohibitions of the word of God: it is the worst disease that has attended this work, most contrary to the spirit and rules of Christianity, and of worse consequences.[74]

This censuring of minister by minister, all within the Puritan New England churches, produced bitter and defensive reactions from established ministers that added fuel to the even larger controversy concerning the emotionalism of the Great Awakening.

In synchrony with the Saturn-Pluto alignment, as early as the autumn of 1740, anti-revivalists began their intense attack on the "emotionalism" and "enthusiasm" witnessed in the events of the Great Awakening.[75] Ever since the turmoil and disruption of the Puritan Revolution of the mid-seventeenth century, many regarded religious "enthusiasm" as a suspect phenomenon responsible for a great portion of that era's disturbing excesses.[76] In this view, religious enthusiasts were seen as fervid believers of a dangerous unorthodoxy who often displayed wild behavior and startling vocal displays and whose claims to experience direct inspiration from God led to dangerous heresies such as antinomianism. Both Anne Hutchinson and George Whitefield were labeled enthusiasts by this definition. As with other aspects of the Great Awakening, it was in New England where "enthusiasm, which seemed to build there to an unmatched crescendo, was seen as the most distinctive, and controversial, of the revival's phenomena."[77] Criticism was directed at the intensely dramatic nature of the preaching styles used by many itinerant preachers as well as the disruptive and suspect behavior displayed by the large revival crowds who heard these sermons and a large percentage of the individual conversion experiences that resulted.

Figure 3 **Outer Planetary Configuration July 8, 1741**

Edwards's Enfield Sermon
Enfield (historical), Massachusetts

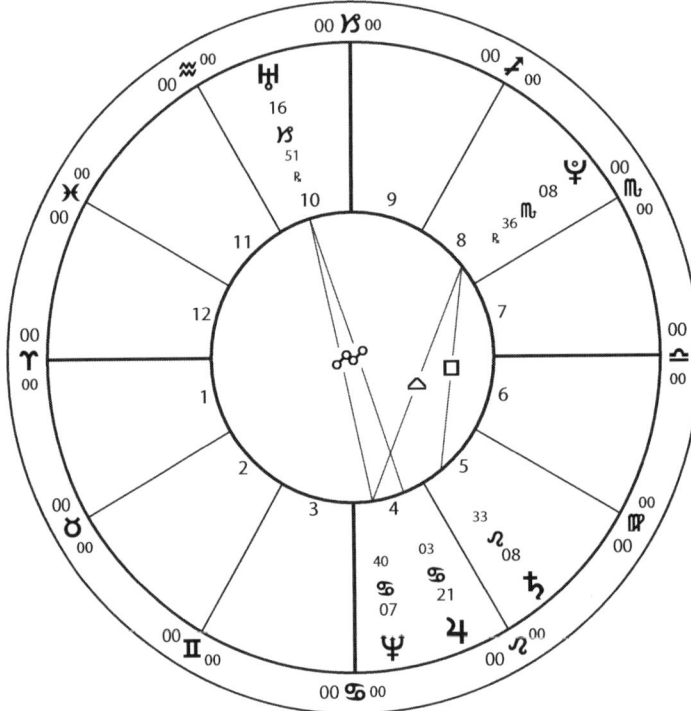

It was during the Great Awakening that this dramatic and enthusiastic preaching style first became widespread in the Anglo-American world, so that by the mid-1740s modern evangelical preaching had "emerged as a distinct form of Christian proclamation."[78] Whitefield was especially known for his "pathetical"—that is, emotional—sermons and the violent emotional response and mass conversions they elicited from the huge crowds. A consummate actor with a charismatic presence and a powerful voice—Benjamin Franklin judged him capable of reaching a crowd of 25,000—rarely did Whitefield's delivery fail to elicit from his audiences not only near perfect silence but also thrills of fear and tears of joy.[79] Known as the "preaching of terror," vivid depictions of the fire and brimstone of hell and the intense sufferings of those who were damned to eternal punishment there were a major component of this new preaching style.

As did many American ministers during this period, Edwards employed the preaching of terror during the Awakening. He did so most effectively in his famous sermon, *Sinners in the Hands of an Angry God,* at Enfield, Massachusetts, in early July 1741, while the Saturn-Pluto square was exactly aligned (see figure 3).[80] The message of Edwards's sermon and its effects on the Enfield congregation that day illustrate quite well the emotionalism criticized by anti-revivalists as well as the many ways in which these phenomena display characteristics of the complex dialectic resulting from the combination of Uranus-Neptune and Saturn-Pluto archetypal themes.

Such "awakening" sermons were intentionally crafted to bring individuals to the most uncomfortable point of terror, to the "brink of hell itself," by painting the most dire picture of the state of their sinful souls and the everlasting torment unregenerate souls are fated to experience—and the resulting cacophony of "shouts and cries" merely demonstrated their success.[81] "Oh sinner!" Edwards warns, "Consider the fearful danger you are in."

> The God that holds you over the pit of hell, much as one holds a spider, or some loathsome insect, over the fire, abhors you, and is dreadfully provoked; his wrath towards you burns like fire; he looks upon you as worthy of nothing else, but to be cast into the fire . . . you are ten thousand times so abominable in his eyes as the most hateful venomous serpent is in ours. . . . and yet 'tis nothing but his hand that holds you from falling into the fire every moment . . . there is no other reason to be given why you han't gone to hell since you have sat here in the house of God, provoking his pure eyes by your sinful wicked manner of attending his solemn worship; yea, there is nothing else that is to be given as a reason why you don't this very moment drop down into hell.[82]

Edwards is not done, however. With his audience staring directly into the fiery pits of eternal torment, Edwards's makes his essential point—that the only thing between such souls and eternal torment is the infinitely merciful hand of a loving God. It is only when a person has reached the point of profound recognition of his or her utter helplessness to avoid the utmost danger that the soul can surrender itself completely to God's mercy and receive the Spirit's saving grace. "And now," Edwards offers,

> you have an extraordinary opportunity, a day wherein Christ has flung the door of mercy wide open, and stands in the door calling and crying with a loud voice to poor sinners . . . many that were lately in the same miserable condition that you are in, are in now an happy state, with their hearts filled with love to him that has loved them and washed them from their sins in his own blood, and rejoicing in hope of the glory of God.[83]

Although hope was truly his message, Edwards's words of terror were so effective that day in Enfield that he likely was unable to present his vision of a loving and merciful God. A witness that day, Stephen Williams, tells us that

> before the sermon was done, there was a great moaning and crying out throughout the whole house. What shall I do to be saved. Oh I am going to Hell. Oh what shall I do for Christ. . . . the shrieks and cries were piercing and amazing.[84]

The outcry was so great that Edwards could not be heard. Though he asked for silence several times, he was unable to finish.[85] Even so, after clergy present that day ministered to those who suffered, sweet release was found by many. "Several souls," Williams continues, "were hopefully wrought upon that night . . . and oh the cheerfulness and pleasantness of their countenances."[86]

Almost formulaic in their development of these two psychological conditions—intense psychological pain followed by a sweet and merciful release in certain knowledge of God's infinite love and forgiveness—such awakening sermons map quite well to the dynamic combination of Saturn-Pluto and Uranus-Neptune archetypal themes.[87] Saturn-Pluto themes, in this context, combine the implacable and immovable judgment of Saturn with the fallen, irrepressibly sinful nature of Pluto to create an intolerably tense condition that, as in an alchemical vessel in which the soul is completely confined and subjected to overwhelming psychological pressures, constrain the individual to wrestle intensely (Pluto) with the fear of eternal punishment decreed by a wrathful God (Saturn) in an unbearable confrontation with utmost terror. In this condition, the unconscious will and instinctual drive associated with

Pluto are completely opposed by the implacable barrier of Saturn, rendering all human efforts completely ineffectual to correct the sinful condition or escape wrathful judgment.

In this impossible struggle, it is only when individuals accept that they are powerless to determine their own fate that the gates open for the soul to be filled and washed clean of terror and sin by the infinite mercy and love of the Spirit. In this second stage of opening and sweet release, we find Neptune's archetypal themes of divine presence, unconditional acceptance, infinite mercy, forgiveness, and all-encompassing love in combination with Uranus' archetypal themes of the sudden breakdown of barriers, an opening up to extraordinary insights, and the thrill of unexpected awakening. When Jupiter's archetypal associations with bountiful optimism, success, and joy are added, the Jupiter-Uranus-Neptune combination especially conveys a sense of successful spiritual conversion, involving a joyful release of fear and terror and a feeling of suddenly, by a stroke of lightning from heaven, being awash in the oceanic illumination of an infinitely good, bountiful, and merciful God. Note also that in this theory of awakening sermons, the "preaching of terror" was thought to be most broadly effective only when the Spirit was present during a particular "season of Grace," when there was an "extraordinary outpouring of the Spirit," which is archetypally quite consistent with phenomena typically found during historical periods when dynamic Uranus-Neptune alignments have formed.[88]

The reaction of the Enfield congregation also serves to illustrate the disruptive nature of the revival events of the Awakening, which were different from all previous revivals in ways that were controversial and consistent with typical Saturn-Pluto associations. Prior to the revivals of the Great Awakening, reactions like that of the Enfield congregation in response to Edwards's *Sinners in the Hands of an Angry God* sermon had never been witnessed in New England. Such outbreaks of loud weeping, moaning, and shrieking—to the point of preventing a preacher from being heard—were not only disruptive but to some observers, because of their highly emotional content, highly suspect. Most suspect were the individual conversions produced during these dramatic and disruptive revival events, which were decidedly more intense and often quite physical, involving uncontrollable episodes of fainting, weeping, shrieking, and shaking.[89] Nothing like these conversions had been seen in New England prior to the Great Awakening. And as described earlier, the question whether or not these events and conversions were truly a work of God formed the central issue of the most intense and divisive theological debates ever experienced within New England Puritanism, between New Lights led by Jonathan Edwards and Old Lights led by Charles Chauncy.[90]

The theological controversy between New and Old Lights was so intense that it shattered any sense of unity or orthodox conformity within New England, forging for the first time within New England Puritanism an entirely new and distinct school of theology, Edwards's New Divinity. Edwards revised Calvinist doctrine, however, in a paradoxical way that also reflects the dynamic combination of Uranus-Neptune and Saturn-Pluto archetypal themes. On the one hand, Edwards primarily intended his revision to confirm and strengthen the three core doctrines of Puritanism—God's absolute sovereignty, the utter depravity of humanity, and predestination—all three fully permeated by Saturn-Pluto themes. But Edwards accomplished this revision of orthodox Calvinistic doctrines by means of new philosophical concepts often associated with more liberal and tolerant Enlightenment movements, such as Arminianism, latitudinarianism, and Unitarianism, which are fully permeated by such typical Uranus-Neptune themes as rational illumination, ecumenism, the spiritual value of self-expression, an individually defined relationship with divinity, and developments toward concepts of religious tolerance and universal salvation.

And by placing the discussion within Lockean terms, Edwards's revision of psychology contributed to the movement by some Old Lights of incorporating the same philosophical and psychological concepts thereby shifting many orthodox Calvinists toward the more liberal positions of Arminianism and Unitarianism. Indeed, Charles Chauncy became one of the leading advocates of liberal New England Arminianism in the late eighteenth century.[91]

Broader Contexts

It is also possible to place the Great Awakening and its many archetypal correlations with the Uranus-Neptune opposition of that period within an even broader geographical context to appreciate just how large a phenomenon this truly was. Something similar took place in several other European countries during the same general period, with major episodes of spiritual awakening and religious revival in Germany, Holland, Switzerland, France, Sweden, Wales, Scotland, and England.[92] A highly significant spiritual awakening or evangelical revival occurred during this period, for example, in Saxony with the emergence of the pietist Moravian movement at the very beginning of the Uranus-Neptune opposition. The thought and practices of the Moravians would play a significant role in the stimulation as well as in the formulation of Methodism.[93] Another important new spiritual impulse was born during this period in Sweden with the spiritual awakening, visions, and writings of Swedenborg, beginning in 1734 and culminating in his intense experiences of 1744–1745.[94] Swedenborg's teachings would play an enormously important role in the thought and works of both Emerson, primarily through the works of Sampson Reed, and William James, through William's father, Henry James, Sr.[95] Hasidism was another highly influential spiritual and religious impulse that emerged during this same period in Poland, starting around 1736 with the mystical and pietistic teachings of Israel ben Eliezer (the Ba'al Shem Tov).[96]

In addition, across the Atlantic from the American Great Awakening but very much connected with it, the English Evangelical Revival flourished in many areas of England, Wales, and Scotland, especially in the wake of the publication of Edwards's *A Faithful Narrative* in 1737, the start of George Whitefield's sensational career as an itinerant preacher in 1737, and the conversion experiences of John and Charles Wesley in 1738.[97] Newspapers in the spring of 1739 reported that on several different occasions the crowds who gathered in the London area to hear Whitefield preach numbered anywhere from ten to thirty thousand.[98] Out of these major periods of revival in both Britain and America, Methodism and modern Evangelicalism were born.

It is also possible to place the Great Awakening within a broader historical context. The Great Awakening represented a transition point in the collective American psyche, perhaps most pronounced in New England where both the preparation for and experience of the Great Awakening were greatest. The preparation lasted more than a century and took the form of a carefully, though imperfectly, constructed and regulated ecclesiastical, political, social, and ideological environment. One important result of that regulated environment was the development in much of New England's population of an enormous and profound psychological expectation that personal regeneration would come through an intense religious conversion experience. During the first century of Puritan New England, that expectation was never met for a large percentage of the population. In profoundly significant ways, the Great Awakening liberated that deep frustration, transformed it, and forever altered the society that contained it—a kind of spiritual emancipation that is archetypally consistent with typical themes found during major Uranus-Neptune alignment periods.

As noted earlier, a central change wrought by the Great Awakening was the democratization of the conversion experience, which combines themes of individual liberty and sudden personal shifts in understanding associated with the Uranus archetype with the religious, spiritual, and unitive mystical experiences associated with the Neptune archetype. Previously thought limited to, or at least more likely to occur in, those properly prepared and in the covenant, the Great Awakening made it clear that this life-changing experience was available to everyone, regardless of church membership and doctrinal preparation. Once the ideological and church affiliation boundaries had been shattered by the Awakening, major periods of revivals continued to sweep through America over the next two centuries, with another major peak of evangelical revivals in the early nineteenth century in the Second Great Awakening and that period's proliferation of new religious, spiritual, utopian, and philosophical movements, including foundational events in the emergence of Unitarianism, Mormonism, Adventism, and Emersonian Transcendentalism. These developments occurred during the very next axial alignment of Uranus and Neptune, the conjunction that began in 1814, was exact in 1821–1822, and ended in 1829.

The theological split that arose from Jonathan Edwards's opposition to Puritan orthodoxy during the Great Awakening resulted in an unprecedented second school of theology, known as the New Divinity. Born as a liberating reaction to orthodox Puritanism, its status quickly became conservative as an even more liberal theological branch developed out of the Great Awakening controversies that would align itself with the Arminianism Edwards sought so ardently to defeat. That liberal trend prepared the ground for both a stronger Anglican presence in what would soon become the United States and the triumph of Unitarianism (1819) in New England in the early nineteenth century, during the same

Uranus-Neptune conjunction (1814–1829) that saw the Second Great Awakening and the birth of Mormonism. Out of that more liberal Unitarian matrix, Emerson's Transcendentalism would emerge to trumpet a new theological and philosophical idealism for New England and America, in which the doctrines of human depravity and predestination were thoroughly denied and replaced with a concept of human dignity and harmony with Nature seen as divine that undermined the very bedrock of Puritan doctrine, the absolute sovereignty of God as a separate, transcendent entity.[99] In Emerson's view, the individual human *is* divine. If you seek conversion, look within—a view that owed much to the profound impression made on Emerson by his encounters with the inner light of Quakerism, which emerged during the Uranus-Neptune conjunction of the mid-seventeenth century.[100]

In another developmental stream, the absorption of the conversion experience into modern schools of psychology can be traced directly to include the theological revision of Jonathan Edwards that developed out of the combination of his earlier synthesis of Enlightenment ideas, his reinterpretation of the conversion experience based on careful observations of the Northampton revival and the Great Awakening, and the fierce theological controversies that arose around those pivotal events. That combination positioned Edwards to envision a new philosophical idealism and an organic psychology of the whole individual that emphasized an integration of the emotions with the understanding. The psychological synthesis of Edwards would find sympathetic resonances in certain aspects of the re-visioning of conversion psychology by William James in his highly influential *The Varieties of Religious Experience*, which was published in 1902 during the very next axial alignment of Uranus and Neptune, the opposition that began in 1899, was exact in 1906–1910, and ended in 1918.

In the *Varieties,* James draws strongly upon Puritan writings regarding the conversion process and the events of the Great Awakening period, especially those of Edwards.[101] While continuing Edwards's emphasis on the empirical examination of individual conversion experiences, James also strove to remove that examination from the domain of theology and philosophy where it had previously resided almost exclusively, producing in the *Varieties* one of the classic, foundational texts for the emerging field of psychology of religion.[102] There is a strong prefiguration in Edwards's psychological model of conversion of the unification model of the conversion process that James presents in *Varieties*, in which the ever-shifting psychological center of a person in the divided, or sick-soul, state moves to a healthier condition, in a process that normally consists "in the straightening out and unifying of the inner self."[103]

Much more could be—and has been—drawn from a comparison of these two prominent American thinkers, but that is outside the scope of this study.[104] William James is born nearly a century after the death of Jonathan Edwards, yet clearly the question of the experience of religious conversion survives in New England even into the early twentieth century when James's *Varieties* was met with critical acclaim and extreme popularity. Edwards's *A Faithful Narrative* (1737) and James's *Varieties* (1902) are similar chords sounding nearly one hundred sixty years apart, both during Uranus-Neptune oppositions, both fundamentally concerned with Uranus-Neptune themes. James's reworking would resonate with the analytical psychology, archetypes, and individuation process developed by C. G. Jung, beginning with the publication of *Symbols of Transformation* in 1912 during the same Uranus-Neptune opposition (1899–1918) in which *Varieties* appeared. The works of James and Jung would produce a host of descendants, including numerous significant contributions in the fields of humanistic psychology, archetypal psychology, and transpersonal psychology.

Transpersonal psychology, in particular, is concerned with the "spiritual dimensions of human nature and existence," especially those "concerns, motivations, experiences, developmental stages . . . modes of being, and other phenomena that include but transcend the sphere of individual personality, self, or ego," which are concerns that are quite consistent with Uranus-Neptune archetypal themes and which received increased elaboration and public attention through numerous publications during and immediately following the most recent Uranus-Neptune conjunction of 1985–2001.[105]

Seasons of Agony and Seasons of Grace

This case study in archetypal historiography has focused on an examination of historical phenomena presented by the New England

Puritan movement looking for patterns of historical correlations with archetypal themes associated with specific cyclical alignments formed by the outer planets. Three major periods of primary importance to the Puritan movement were identified: the emergence of the Puritan movement in Elizabethan England; the Half-Way Covenant Crisis period; and the Great Awakening. Although other significant, long-term outer-planetary alignments were involved, all three periods occurred in synchrony with consecutive axial alignments of the Uranus-Neptune cycle. The Puritan movement emerged in 1566 during the first Uranus-Neptune axial alignment, the opposition of 1556–1574. The Half-Way Covenant controversy formed when revisionist ministers organized to oppose the foundational doctrinal position of the New England Way, published as the *Cambridge Platform* in 1648, during the very next Uranus-Neptune axial alignment, the conjunction of 1643–1658. And the events of the Great Awakening took place during the very next axial alignment, the Uranus-Neptune opposition of 1728–1746. Moreover, either the initiating or the central, defining event of each of these three periods took place during those years in which the Uranus-Neptune alignment was exact or near exact: Puritanism in 1566 while the opposition was exactly aligned (1563–1566); the Cambridge Synod in 1648 while that conjunction was exactly aligned (1648–1650); and the Northampton revivals, Edwards's foundational descriptions of those events, and the birth of Methodism in 1734–1738 while that opposition was exactly aligned (1734–1738).

The tight coupling of these major historical periods with the Uranus-Neptune cycle seems especially resonant with the explanation offered by evangelical Puritans for the completely unprecedented series of massive revivals and thousands of individual conversion experiences of the Great Awakening. These surprising events took place because God granted a rare season of grace when an extraordinary outpouring of Spirit possessed individual souls to regenerate them, changing lives forever. An expanded notion of such seasons of grace or great awakenings as periods of significant religious and spiritual revival, when something extraordinary does take place to transform individual lives, revive religious institutions, revitalize a culture, and transform its vision, is a compelling one and not only for evangelical Protestants. In *Revivals, Awakenings, and Reform,* William McLoughlin distinguishes great awakenings from "the religious revivalism that accompanies them."[106] While Protestant revivalism is generally an important component of all great awakenings in America, he defines great awakenings as much broader phenomena. For McLoughlin, great awakenings are

> periods of cultural revitalization that begin in a general crisis of beliefs and values and extend over a period of a generation or so, during which time a profound reorientation in beliefs and values takes place. Revivals alter the lives of individuals; awakenings alter the world view of a whole people or culture.[107]

Combining this general concept of great awakenings as major periods of cultural revitalization with the general historical "agreement that widespread expressions of religious concern have recurred periodically in American history," McLoughlin discerns five American Great Awakenings that in each case led to significant cultural transformations (see table 2).[108]

Table 2   **McLoughlin's Five American Great Awakenings**

Awakening	Period
Puritan Awakening	Early to mid-17th century
First Great Awakening	Early to mid-18th century
Second Great Awakening	Early 19th century
Third Great Awakening	Late 19th and early 20th centuries
Fourth Great Awakening	Final decades of the 20th century

Using another approach, Richard Tarnas has examined the archetypal patterns presented by a wide range of cultural phenomena over many centuries of Western history. Tarnas finds historical periods during which dynamic Uranus-Neptune alignments have formed to be "characterized . . . by pervasive transformations of a culture's underlying vision," including

> widespread spiritual awakenings, the birth of new religious movements, cultural renaissances, the emergence of new philosophical perspectives, rebirths of idealism, sudden shifts in a culture's cosmological and metaphysical vision, rapid collective changes in psychological understanding and interior sensibility, certain forms of

scientific paradigm shifts, new utopian social visions and movements, and epochal shifts in a culture's artistic imagination.[109]

These two summaries have much in common. For example, McLoughlin describes the cultural transformation that takes place during great awakenings as the "alteration of the world view of an entire culture," while for Tarnas dynamic alignments of Uranus-Neptune often correlate with "the pervasive transformation of that culture's underlying vision."

As we have seen, the first Great Awakening displays phenomena that are highly consistent with both summaries.[110] Most obviously, the Great Awakening presents a pattern of widespread spiritual awakenings and the birth of new religious movements that transformed the lives and thought of an enormous number of individuals in American culture during the 1730s, 1740s, and beyond. And in Jonathan Edwards's philosophical and theological revision, we also find a new formulation of philosophical idealism and a new psychological understanding of interiority that is both representative of and a catalyst for a much larger cultural shift in philosophical and religious perspectives.

The Great Awakening as an enormously important cultural shift is precisely the conclusion reached by historians of the period. Briefly presented earlier but worth returning to at this point, Heimert and Miller, for example, consider the effects of the Great Awakening to be fundamentally significant. To them "the Awakening clearly began a new era, not merely of American Protestantism, but in the evolution of the American mind . . . a watershed in American development."[111] It was "the end of the reign over the New England and American mind of a European and scholastic conception of an authority put over men because men were incapable of recognizing their own welfare."[112] "After 1750," they continue,

> we are in a "modern period," whereas before that, and down to the very outburst, the intellectual world is still medieval, scholastic, static, authoritarian. The Awakening . . . marked America's final break with the Middle Ages and her entry into a new intellectual age in the church and in society. . . . The Great Awakening thus stands as a major example of that most elusive of phenomena: a turning point, a "crisis," in the history of American civilization.[113]

In religious terms, Heimert and Miller find that the Awakening "introduced a profound shift in the very character, the perspective and focus, of religious thought and discourse."[114] Out of the debates of that period, "came the reorientations in religious thought that clearly mark the Great Awakening as the 'birth-pangs' of a new epoch in the history of American Protestantism."[115] In a broader sense, Sydney Ahlstrom asserts that the "social and political consequences of the Awakening are so important and so widely ramified that they can be discussed only in the context of the country's ongoing experience."[116] And McLoughlin sees the American Revolution as "the secular fulfillment of the religious ideals of the First Great Awakening," concluding that "Our Revolution" came after the First Great Awakening on American soil had made the thirteen colonies into a cohesive unit *(e pluribus unum),* had given them a sense of unique nationality, and had inspired them with the belief that they were, "and of right ought to be," a free and independent people.[117]

Examined archetypally, Tarnas found a wealth of similar phenomena during dynamic Uranus-Neptune alignments, especially their axial alignments (conjunctions and oppositions).[118] Since the Reformation, which is the starting point of this case study, there have been six axial alignments of Uranus and Neptune. Table 3 lists those six alignments along with a summary of the more significant historical correlations for each of these periods that are relevant to this case study. As might be expected, McLoughlin's five Great Awakenings are readily integrated into this scheme, with each Great Awakening corresponding in sequence with consecutive axial alignments of Uranus and Neptune. In light of this remarkable pattern of archetypal correlations of axial Uranus-Neptune alignments with major historical periods of religious revival, great awakenings, and cultural transformation, the Puritan notion of seasons of grace in which extraordinary outpourings of Spirit occur is archetypally consistent with core themes of the Uranus-Neptune complex.

Table 3   **The Uranus-Neptune Cycle and Major Cultural Awakenings**

Alignment	Dates	Events
Opposition	1556–1574	Elizabethan Religious Settlement • Thirty-nine Articles (1563) • Puritan movement emerges (1566).
Conjunction	1643–1658	Puritan Awakening • Half-Way Covenant crisis (1645–1662) • *Cambridge Platform* (1648) defines New England Way. • England's Puritan Revolution • Westminster Confession (1646) • Birth of Quakerism • Proliferation of radical religious sects in England
Opposition	1728–1746	First Great Awakening • Transition in America from medieval to modern world view • Orthodox Puritan hegemony broken • Democratization of the conversion experience • Jonathan Edwards revises Puritan theology and conversion psychology. • Evangelical Revival in England, Scotland, Wales • Evangelical piety movements in Holland, Germany, France • Birth of Methodism, modern Evangelicalism, Moravianism, Hasidism, and Swedenborgianism

Conjunction	1814–1829	Second Great Awakening
		• Arminianism triumphs in New England:
		  – Emergence of Unitarianism (1819)
		  – Decline and liberalization of orthodox Puritanism
		• Roots of Emersonian Transcendentalism
		• Birth of Mormonism, Adventism
		• Major period of Methodist expansion
		• Social reform movements gain momentum
		• Proliferation of social utopian communities

Opposition	1899–1918	Third Great Awakening • William James publishes *The Varieties of Religious Experience* (1902) advancing the process of interpreting the conversion experience psychologically. • Jung publishes *Symbols of Transformation* (1912) • Max Weber publishes *The Protestant Ethic and the "Spirit" of Capitalism* (1904–1905) • Birth of Pentecostalism • Social Gospel movement born.

Table 3   **The Uranus-Neptune Cycle and Major Cultural Awakenings**

Conjunction	1985–2001	Fourth Great Awakening
		• Pentecostalism shows explosive world-wide growth.
		• Charismatic Christian movement expands.
		• New Age spirituality proliferates primarily in America and other English speaking countries, as well as in Holland, Germany, and France.
		• Transpersonal psychology provides new models of psychological rebirth and spiritual experiences.

Furthermore, when the phenomena of each of these three major periods of Puritan history were examined in greater detail, another notable set of correlations were revealed with those archetypal complexes associated with other specific planetary alignments of the period. While notable patterns of correlations were discovered for the Neptune-Pluto, Uranus-Pluto, Saturn-Uranus, and Jupiter-Uranus cycles, possibly the most distinctive pattern was found in the remarkable sequence of correlations during dynamic alignments of the Saturn-Pluto cycle, when major times of controversy producing specific types of crises formed with remarkable consistency. For convenience, many of the correlations in this expanded sequence are again presented in table 4.

Table 4   **The Saturn-Pluto Cycle and Major Puritan Events**

Alignment	Years	Events
Conjunction	1516–1519	Birth of Reformation, Luther's Ninety-five Theses (1517).
Opposition	1533–1536	Henry VIII's Act of Supremacy (1534) St. Ignatius of Loyola publishes *Spiritual Exercises* (1535). Calvin publishes *Institutes of the Christian Religion* (1536).
Conjunction	1551–1554	England's Edward VI dies. Reign of Mary I begins period of Protestant persecution, execution, and exile (1553).
Opposition	1565–1569	Puritanism as a unified movement emerges (1566). Major ecclesiastic publication controversy in England
Opposition	1630–1633	Great Puritan Migration begins (1631).
Square	1640–1642	First English Civil War begins (1642). Great Puritan Migration slows.

Beyond a Disenchanted Cosmology

Table 4   **The Saturn-Pluto Cycle and Major Puritan Events**

Conjunction	1647–1650	Cambridge Synod promulgates *Cambridge Platform* (1648). Pride's Purge; second English Civil War begins (1648). Charles I executed (1649).
Square	1654–1656	Connecticut Half-Way crisis (1656) Ministerial Assembly called (1656).
Opposition	1662–1666	Half-Way Covenant publishing controversy (1663) King's Commissioners arrive to assume control of New England; suppress religious debate (1664).
Conjunction	1679–1682	Reforming Synod, Mass Covenant Renewals (1679–1680)
Square	1740–1742	Great Awakening controversy (1740–1741)

Another of the more significant findings of this study is that the central defining characteristics of the Puritan movement—its fundamental goals and concerns, and the essential dynamics and processes of its most important, specific events—were also found to be highly consistent with the archetypal dynamics contained within a synthesis of themes associated with the Uranus-Neptune and Saturn-Pluto complexes. The evidence presented here suggests that the character of the Puritan movement reflects in coherent and consistent ways the rare astronomical configuration of the four outermost planets in 1566, the year the Puritan movement first emerged in Elizabethan England (see figure 4).

The evidence also shows that the Puritan movement responded in ways that are highly characteristic of this natal configuration during subsequent periods when similar configurations of planetary alignments have formed. Such characteristic responses were found especially in the significant events of the two major periods after the birth of the Puritan movement, including the Half-Way Covenant controversy during the Cambridge Synod and the many controversies of the Great Awakening period. Both took place when the Uranus-Neptune and Saturn-Pluto cycles had again formed dynamic alignments simultaneously, suggesting that some kind of archetypal resonance is occurring within the collective Puritan psyche during periods of similar planetary configurations. Remarkably, these three periods were the only ones between 1500 and 1800 when Uranus and Neptune were in axial alignment and Saturn and Pluto were in quadrature alignment simultaneously.

Figure 4 **Planetary Configuration during Birth of Puritan Movement**

August 25, 1566
London, England

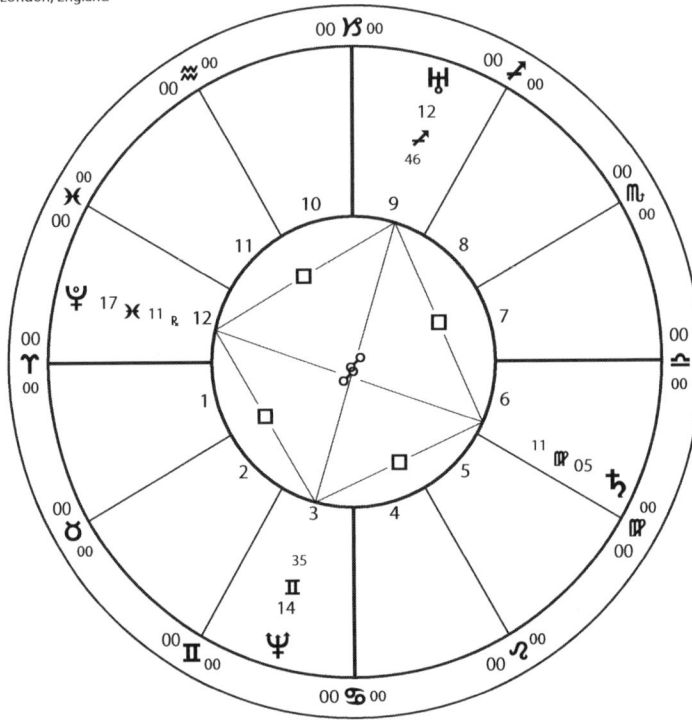

The evidence presented in this case study and elsewhere by Richard Tarnas and other researchers seems to indicate that something like a concept of seasons within cultural history may be appropriate and that these seasons reflect in some profound way both a synchrony and an archetypal resonance with the dynamic alignments formed by the outer planets. This archetypal synchronous reflection seems to apply not only for larger periods of Great Awakenings, so that in some sense there are indeed seasons of grace in which an extraordinary outpouring of Spirit might take place. But it also seems to hold for shorter periods of personal and cultural agony, conflict, and crisis, in which psychological battle, both external and internal, take place. The Puritan movement demonstrates a profoundly complex pattern of phenomena containing both kinds of cultural seasons that are closely synchronized with dynamic Uranus-Neptune and Saturn-

Pluto alignments and that are deeply resonant with the relevant archetypal complexes associated with those alignments.

Such thoughts raise a host of important questions and issues, of course. The most important question might be: How is it possible that such a relationship exists between human collective events and planetary configurations? The mere postulation of any such connection between human and astronomical events seems to fly in the face of current definitions of causation and raises the ever thorny issue of fate. But if such a relationship does indeed exist, by which we are able to discern these kinds of archetypal resonances and patterns in history, then fundamental aspects of our current understanding about the nature of reality may be rightly questioned.[119]

Notes

1. On this revival theory, see Frank Lambert, *Inventing the "Great Awakening"* (Princeton: Princeton University Press, 1999), 43–50.
2. Lambert, *Inventing the "Great Awakening,"* 46.
3. Lambert, *Inventing the "Great Awakening,"* 46–47.
4. Lambert, *Inventing the "Great Awakening,"* 48, 53.
5. Lambert, *Inventing the "Great Awakening,"* 44.
6. Winslow, *Jonathan Edwards*, 159.
7. George Marsden, *Jonathan Edwards: A Life* (New Haven: Yale University Press, 2003), 158–159.
8. Marsden, *Jonathan Edwards*, 159.
9. Ahlstrom, *Religious History*, 301.
10. Marsden, *Jonathan Edwards*, 159.
11. Marsden, *Jonathan Edwards*, 160.
12. Heimert and Miller, *Great Awakening*, xviii.
13. Quoted in Ahlstrom, *Religious History*, 301–302.
14. Mark A. Noll, *The Rise of Evangelicalism: The Age of Edwards, Whitefield, and the Wesleys* (Downers Grove, IL: InterVarsity Press, 2003), 95–99.
15. Lambert, *Inventing the "Great Awakening,"* 101.
16. Bremer, *Puritan Experiment*, 229; and Marsden, *Jonathan Edwards*, 202–205.
17. Heimert and Miller, *Great Awakening*, xiii.
18. Lambert, *Inventing the "Great Awakening,"* 21–25.

19. Ahlstrom, *Religious History*, 286.

20. Ahlstrom, *Religious History*, 286.

21. Edward L. Glaeser, "Reinventing Boston: 1640–2003," Harvard Institute of Economic Research, Discussion Paper Number 2017 (Cambridge, MA: Harvard University, September 2003), http://www.economics.harvard.edu/pub/hier/2003/HIER2017.pdf.

22. Marsden, *Jonathan Edwards*, 219.

23. Marsden, *Jonathan Edwards*, 216.

24. Ahlstrom, *Religious History*, 286–287.

25. For the content of this paragraph, including quotations, see Ahlstrom, *Religious History*, 286–287.

26. Ahlstrom, *Religious History*, 285.

27. Ahlstrom, *Religious History*, 287.

28. Heimert and Miller, *Great Awakening*, xxxviii.

29. A nearly identical division occurred during this same period in the Presbyterian Churches of the Middle Colonies, where the two positions were known as Old Sides and New Sides. Lambert, *Inventing the "Great Awakening,"* 132.

30. Heimert and Miller, *Great Awakening*, xxxix.

31. On the life of Jonathan Edwards, I follow primarily Marsden, *Jonathan Edwards;* Winslow, *Jonathan Edwards;* Ahlstrom, *Religious History*, 280–313; and Perry Miller, *Jonathan Edwards* (New York: Meridian Books, 1959). For valuable shorter biographical presentations, see Flower and Murphey, *History*, 137–142; and Marty, *Pilgrims*, 113–117.

32. Marsden, *Jonathan Edwards*, 59.

33. See, for example, Flower and Murphey, *History*, 137, who rank Edwards as "the greatest American theologian and the greatest American philosopher before the Civil War."

34. On Jonathan Edwards' philosophical and theological synthesis, I draw primarily on Marsden, *Jonathan Edwards*, especially 59–81; Flower and Murphey, *History*, 142–183; and Ahlstrom, *Religious History*, 280–313.

35. Flower and Murphey, *History*, 142.

36. Copleston, "Newton," *A History of Philosophy*, 5:155.

37. No evidence has been found for a direct influence of Berkeley on Edwards. George Berkeley's most influential work outlining his immaterialism (or subjective idealism) was his *Treatise Concerning the Principles of Human Understanding*, published in 1710 (repr., New York: Cosimo, 2005). Although the overall philosophical positions of the systems of Edwards and Berkeley are very similar, according to Flower and Murphey (*History*, 142), the evidence from Edwards' notebooks shows his position to have substantial differences with Berkeley's and presents absolutely no mention of Berkeley in what are otherwise meticulously documented and referenced documents.

38. This summary of Locke's epistemology is based on Copleston, "Locke," *History of Philosophy*, 5:67–122; and William Uzgalis, "John Locke," (2007) in *Stanford Encyclopedia of Philosophy*, http://plato.stanford.edu/entries/locke/.

39. Flower and Murphey, *History*, 144.

40. Edwards, "The Mind," quoted in Flower and Murphey, *History*, 144.

41. Copleston, "Locke," *History of Philosophy*, 5:81–86.

42. Copleston, "Locke," *History of Philosophy*, 5:82, 84–85.

43. Edwards, "Mind," quoted in Flower and Murphey, *History*, 144.

44. Edwards, "Mind," quoted in Flower and Murphey, *History*, 144.

45. Edwards, "Notes on Natural Science," quoted in Flower and Murphey, *History*, 144.

46. On Edwards' trinitarian formulation, I follow Flower and Murphey, *History*, 156–157.

47. Charles Chauncy, *Seasonable Thoughts on the State of Religion* (1743), quoted in Ahlstrom, *Religious History*, 302–303.

48. Heimert and Miller, *Great Awakening*, xl.

49. Edwards, *Some Thoughts Concerning the Present Revival of Religion in New England* (1742), in Heimert and Miller, *Great Awakening*, 264–265.

50. Edwards, *Concerning the Present Revival*, in Heimert and Miller, *Great Awakening*, 265.

51. Heimert and Miller, *Great Awakening*, xl.

52. Edwards, *Concerning the Present Revival*, in Heimert and Miller, *Great Awakening*, 265.

53. Edwards, *Concerning the Present Revival*, in Heimert and Miller, *Great Awakening*, 276.

54. Ahlstrom, *Religious History*, 302–303.

55. Edwards, *A Treatise Concerning Religious Affections* (1746), quoted in Ahlstrom, *Religious History*, 302–303.

56. See, for example, Jorge Ferrer, *Revisioning Transpersonal Psychology: A Participatory Vision of Human Spirituality* (Albany: State University of New York Press, 2002), 5–6. On the influence Emerson and James have had on humanistic and transpersonal psychology, see Wouter J. Hanegraaff, *New Age Religion and Western Culture: Esotericism in the Mirror of Secular Thought* (Albany: State University of New York Press, 1998), especially 50–52 and 490–496.

57. See, for example, Ahlstrom, *Religious History*, 294.

58. Heimert and Miller, *Great Awakening*, xli.

59. Ahlstrom, *Religious History*, 288–289.

60. Bushman, *From Puritan to Yankee*, 187.

61. See, for example, Ahlstrom, *Religious History*, 294.

62. Heimert and Miller, *Great Awakening*, xv–xvi.

63. Ahlstrom, *Religious History*, 294. See also Noll, *Rise of Evangelicalism*, 136–154, who offers an entire chapter as an insightful overview of the many legitimate factors—including economic, social, ecclesiastical, international, and psychological—that historians have emphasized in their attempts to explain the Great Awakening and the rise of evangelicalism during the 1730s and 1740s in Britain and America; and

McLoughlin, *Revivals, Awakenings, and Reform,* 52–59, who considers five broad and distinct categories of explanations.

64. See especially Noll, chapter 3, "Revival, 1734–1738," *Rise of Evangelicalism,* 76–99.

65. Lambert, *Inventing the "Great Awakening,"* 251–253, who indicates that in 1744 and 1745 most revivalists had concluded that the outpouring of the Spirit had died down in both the American colonies and across the Atlantic.

66. Tarnas, *Cosmos and Psyche,* 356.

67. See Tarnas, "Cycles of Creativity and Expansion," *Cosmos and Psyche,* 289–351; Edwards, *Faithful Narrative,* quoted in Marsden, *Jonathan Edwards,* 159.

68. Marsden, *Jonathan Edwards,* 205–206.

69. Ahlstrom, *Religious History,* 286.

70. Lambert, *Inventing the "Great Awakening,"* 218; Marsden, *Jonathan Edwards,* 210–211.

71. Marsden, *Jonathan Edwards,* 210.

72. Lambert, *Inventing the "Great Awakening,"* 58, 192, 240; and Richard L. Bushman, ed., *The Great Awakening: Documents on the Revival of Religion, 1740–1745* (1970; repr., Chapel Hill: University of North Carolina Press, 1989), 85–87.

73. Gilbert Tennent, *The Danger of an Unconverted Ministry* (1740), in *Great Awakening,* Bushman, 90.

74. Edwards, *Concerning the Present Revival,* in *Great Awakening,* Heimert and Miller, 282–283.

75. Marsden, *Jonathan Edwards,* 215.

76. See, for example, Noll, *Rise of Evangelicalism,* 55–56.

77. Heimert and Miller, *Great Awakening,* xxxviii.

78. Noll, *Rise of Evangelicalism,* 132–134; and Lambert, *Inventing the "Great Awakening,"* 97.

79. Lambert, *Inventing the "Great Awakening,"* 96–98; Marsden, *Jonathan Edwards,* 206.

80. Marsden, *Jonathan Edwards,* 219, who has Edwards delivering this sermon at Enfield on July 8, 1741 (New Style).

81. Lambert, *Inventing the "Great Awakening,"* 98.

82. Jonathan Edwards, *Sinners in the Hands of an Angry God* (1741), in *A Jonathan Edwards Reader,* ed. John E. Smith, Harry S. Stout, and Kenneth P. Minkema (New Haven: Yale University Press, 1995), 97–98.

83. Edwards, *Sinners,* in *Edwards Reader,* Smith et al., 103.

84. Marsden, *Jonathan Edwards,* 220, whose account is taken from Stephen William's unpublished diary, Storrs Memorial Library, Longmeadow, MA (*Jonathan Edwards,* 549, n15–16).

85. Marsden, *Jonathan Edwards,* 221.

86. Marsden, *Jonathan Edwards,* 221.

87. On effective preaching and awakening sermons, see Lambert, *Inventing the "Great Awakening,"* 46–50.

88. Lambert, *Inventing the "Great Awakening,"* 45–50.

89. Ahlstrom, *Religious History,* 286; Marsden, *Jonathan Edwards,* 218.

90. Although it is certainly true that other intense and divisive debates took place within New England Puritanism, notably in those controversies surrounding Roger Williams, Anne Hutchinson, and the Half-Way Covenant, none of these earlier debates produced a theological division among New England Puritan clergy and congregations as profound or as widespread as did the Great Awakening controversies. which for the first time resulted in two entirely separate schools of theology and severely fractured New England church uniformity.

91. Conrad Wright, *The Beginnings of Unitarianism in America* (Boston: Starr King Press, 1955), 135–160; Alan Heimert, *Religion and the American Mind: From the Great Awakening to the Revolution* (Cambridge, MA: Harvard University Press, 1966), note on 109–110.

92. Heimert and Miller, Great Awakening, xv; Noll, Rise of Evangelicalism, 68–135.

93. Noll, *Rise of Evangelicalism,* 68, 84–86, 93–99.

94. Ernst Benz, *Emanuel Swedenborg: Visionary Savant in the Age of Reason,* 2nd ed., trans. Nicholas Goodrick-Clarke (West Chester, PA: Swedenborg Foundation, 2002), 151–200; and Michael Stanley, ed., *Emanuel Swedenborg* (Berkeley, CA: North Atlantic Books, 2003), 16–25.

95. Robert D. Richardson, Jr., *Emerson: The Mind on Fire* (Berkeley and Los Angeles: University of California Press, 1995), 6–63; Frederic Harold Young, *The Philosophy of Henry James, Sr.* (New York: Bookman Associates, 1951), 29–65.

96. Gershom Scholem, *Major Trends in Jewish Mysticism,* foreword by Robert Alter (1941; repr., New York: Schocken Books, 1995), 325–350.

97. Noll, *Rise of Evangelicalism,* 86–99; Lambert, *Inventing the "Great Awakening,"* 102–110.

98. Lambert, *Inventing the "Great Awakening,"* 105.

99. Ahlstrom, *Religious History,* 600–606, who describes Emerson as "the theologian of something we may almost term 'the American religion'" (605), which he later describes as Harmonial Religion (1019–1036). During this period from 1817 through 1829, Emerson was at Harvard, studied intensively with Ellery Channing (who provided the founding statement for Unitarianism in America in 1819), and absorbed the works of Plato and the Neoplatonic system of Swedenborg, as elaborated by Sampson Reed, among other important influences. See Richardson, *Emerson,* 6–83.

100. Richardson, *Emerson,* 78, 125, 157–163.

101. Gerald E. Myers, *William James: His Life and Thought* (New Haven: Yale University Press, 1986), 456. Numerous references to and quotations from the works of Edwards may be found in two sections on "Conversion" (Lectures XI and X), James, *Varieties,* 189–258.

102. David M. Wulff, *Psychology of Religion: Classic and Contemporary Views* (New York: John Wiley and Sons, 1991), 11–12, 498–501.

103. James, *Varieties,* 170.

104.See, for example, William A. Clebsch, *American Religious Thought: A History*, foreword by Martin E. Marty (Chicago: University of Chicago Press, 1973), who examines the many connections among the thought of Edwards, Emerson, and James as the primary expositors of a uniquely American vein of religious thought that is focused on the harmony between human beings and the world and universe we live in.

105.Ferrer, *Revisioning Transpersonal Theory,* 5. See Tarnas, *Cosmos and Psyche,* 401–408 on the ways that Uranus-Neptune themes in the works of James and Jung prepared the ground for both "archetypal and transpersonal psychology, two of the most vital currents to emerge from the wellspring of depth psychology in the past several decades."

106.McLoughlin, *Revivals, Awakenings, and Reform*, xiii.

107.McLoughlin, *Revivals, Awakenings, and Reform*, xiii.

108.Published in 1978, McLoughlin gives rough periods for the dates of these great awakenings, to which he would not want "to be held strictly." McLoughlin, *Revivals, Awakenings, and Reform,* 8, 10–11. The dates he gives are as follows: "the Puritan Awakening, 1610–1640; the First Great Awakening (in America), 1730–1760; the Second Great Awakening, 1800–1830; the Third Great Awakening, 1890–1920; and the Fourth Great Awakening, 1960–90(?)" (McLoughlin, *Revivals, Awakenings, and Reform,* 10–11). McLoughlin's theory of Great Awakenings deserves a more careful and detailed examination from the perspective of archetypal historiography than is possible here. In significant ways, the evidence presented in this case study and by Tarnas in *Cosmos and Psyche* supports McLoughlin's general theory of Great Awakenings while also pointing to the need for specific adjustments. The most important differences between the body of archetypal evidence and the dates given by McLoughlin in 1978 concern the Puritan Awakening and the Fourth Great Awakening. The archetypal and historical evidence appears to support dates for the Puritan Awakening between 1630 and 1660. And from the perspective of archetypal historiography, McLoughlin's dates for the Fourth Great Awakening clearly include the phenomena associated with the Uranus-Pluto conjunction of 1960–1972, and should be expanded to include the phenomena associated with the Uranus-Neptune conjunction of 1985–2001. For the larger discussion of Uranus-Pluto correlations, including specific sections on the 1960s, see Tarnas, *Cosmos and Psyche,* 141–205. For a larger discussion of Uranus-Neptune correlations, see Tarnas, *Cosmos and Psyche,* 353–451, including a specific section on the late twentieth century, 419–451.

109.Tarnas, *Cosmos and Psyche,* 356.

110.Tarnas includes America's first Great Awakening and the significant cultural shifts that resulted from it as among the important correlations for the Uranus-Neptune opposition of 1728–1746. See *Cosmos and Psyche,* 372.

111.Heimert and Miller, *Great Awakening,* xiv.

112.Heimert and Miller, *Great Awakening,* lxi.

113.Heimert and Miller, *Great Awakening,* xiv–xv.

114.Heimert and Miller, *Great Awakening,* xliii.

115.Heimert and Miller, *Great Awakening,* xxxviii.

116.Ahlstrom, *Religious History,* 294. For more recent, in-depth assessments that differ significantly in intent while reaching conclusions quite similar to that of Ahlstrom, see Lambert, *Inventing the "Great Awakening";* and Noll, *Rise of Evangelicalism.*

117.McLoughlin, *Revivals, Awakenings, and Reform,* 97, 1.

118. Tarnas, "Awakenings of Spirit and Soul," *Cosmos and Psyche,* 353–451.

119. The theoretical implications of this case study are explored further in chapter 9 of my dissertation, "Seasons of Agony and Grace," 338–373.

Bibliography

Ahlstrom, Sydney E. *A Religious History of the American People.* New Haven: Yale University Press, 1974.

Bainton, Roland H. *The Reformation of the Sixteenth Century.* Forward by Jaroslav Pelikan. Boston: Beacon Press, 1985.

Barzun, Jacques. *From Dawn to Decadence: 500 Years of Western Cultural Life 1500 to the Present.* 2000. Reprint, New York: Perennial / HarperCollins Publishers, 2001.

Bremer, Francis J. *The Puritan Experiment: New England Society from Bradford to Edwards.* Hanover, NH: University Press of New England, 1976.

Bushman, Richard L. *From Puritan to Yankee: Character and the Social Order in Connecticut, 1690–1765.* Cambridge, MA: Harvard University Press, 1967.

―――, ed. *The Great Awakening: Documents on the Revival of Religion, 1740–1745.* 1970. Reprint, Chapel Hill: University of North Carolina Press, 1989.

Clebsch, William A. *American Religious Thought: A History.* Foreword by Martin E. Marty. Chicago: University of Chicago Press, 1973.

Cohen, Charles Lloyd. *God's Caress: The Psychology of Puritan Religious Experience.* New York: Oxford University Press, 1986.

Collinson, Patrick. *The Elizabethan Puritan Movement.* 1967. Reprint, New York: Oxford University Press, 1990.

―――. *The Reformation: A History.* New York: Modern Library, 2004.

Cox, Harvey. *Fire from Heaven: The Rise of Pentecostal Spirituality and the Reshaping of Religion in the Twenty-first Century.* 1995. Reprint, Cambridge, MA: Da Capo Press, 2001.

Ebertin, Reinhold. *The Combination of Stellar Influences.* Tempe, AZ: American Federation of Astrologers, 1997.

———. *Transits.* Tempe, AZ: American Federation of Astrologers, 1995.

Edwards, Jonathan. *Sinners in the Hands of an Angry God.* 1741. In *A Jonathan Edwards Reader.* Edited by John E. Smith, Harry S. Stout, and Kenneth P. Minkema. New Haven: Yale University Press, 1995.

———. *Some Thoughts Concerning the Present Revival of Religion in New England.* 1742. In *The Great Awakening: Documents Illustrating the Crisis and Its Consequences.* Edited by Alan Heimert and Perry Miller. New York: Bobbs-Merrill, 1967.

Flower, Elizabeth, and Murray G. Murphey. *A History of Philosophy in America.* 2 vols. New York: Capricorn Books / G. P. Putnam's Sons, 1977.

Greene, Liz. *The Astrological Neptune and the Quest for Redemption.* York Beach, ME: Samuel Weiser, 1996.

———. *The Astrology of Fate.* York Beach, ME: Samuel Weiser, 1995.

———. *Saturn: A New Look at an Old Devil.* York Beach, ME: Samuel Weiser, 1976.

Greene, Liz, and Howard Sasportas. *The Development of the Personality.* Seminars in Psychological Astrology, vol. 1. York Beach, ME: Samuel Weiser, 1987.

———. *Dynamics of the Unconscious.* Seminars in Psychological Astrology, vol. 2. York Beach, ME: Samuel Weiser, 1988.

Hand, Robert. *Horoscope Symbols.* Atglen, PA: Whitford Press, 1981.

———. *Planets in Transit: Life Cycles for Living.* Atglen, PA: Whitford Press, 1976.

Heimert, Alan. *Religion and the American Mind: From the Great Awakening to the Revolution.* Cambridge, MA: Harvard University Press, 1966.

Heimert, Alan, and Perry Miller, eds. *The Great Awakening: Documents Illustrating the Crisis and Its Consequences.* American Heritage Series.

Beyond a Disenchanted Cosmology

Leonard W. Levy and Alfred Young, general eds. New York: Bobbs-Merrill Company, 1967.

Hill, Christopher. *The World Turned Upside Down: Radical Ideas During the English Revolution.* 1975. Reprint, London: Penguin Books, 1991.

James, William. *The Varieties of Religious Experience: A Study in Human Nature.* 1902. Reprint, edited and introduced by Marty E. Martins. New York: Penguin Books, 1982.

Kishlansky, Mark. *A Monarchy Transformed: Britain 1603–1714.* London: Penguin Books, 1996.

Lambert, Frank. *Inventing the "Great Awakening."* Princeton: Princeton University Press, 1999.

Marsden, George. *Jonathan Edwards: A Life.* New Haven: Yale University Press, 2003.

Marty, Martin E. *Pilgrims in Their Own Land: 500 Years of Religion in America.* Boston: Little, Brown and Company, 1984.

McDermott, Robert A. "The Spiritual Mission of America," Revision, 16. no. 1, (Summer 1993): 15–25.

McLoughlin, William G. *Revivals, Awakenings, and Reform: An Essay on Religion and Social Change in America, 1607–1977.* 1978. Reprint, Chicago and London: University of Chicago Press, 1980.

Miller, Perry. *Jonathan Edwards.* New York: Meridian Books, 1959.

———. *The New England Mind: From Colony to Province.* 1953. Reprint, Cambridge, MA: Belknap Press / Harvard University Press, 1998.

———. *The New England Mind: The Seventeenth Century.* 1939. Reprint, Boston, MA: Beacon Press, 1961.

Morgan, Edmund S. *Visible Saints: The History of a Puritan Idea.* 1963. Reprint, Ithaca, NY and London: Cornell University Press, 1965.

Myers, Gerald E. *William James: His Life and Thought.* New Haven: Yale University Press, 1986.

Noll, Mark A. *The Rise of Evangelicalism: The Age of Edwards, Whitefield, and the Wesleys.* Downers Grove, IL: InterVarsity Press, 2003.

O'Neal, Rod. "Seasons of Agony and Grace: An Archetypal History of New England Puritanism." PhD diss., California Institute of Integral Studies, 2008.

Pope, Robert. *The Half-Way Covenant: Church Membership in Puritan New England.* Eugene, Oregon: Wipf and Stock Publishers, 1969.

Richardson, Robert D., Jr. *Emerson: The Mind on Fire.* Berkeley and Los Angeles: University of California Press, 1995.

Rudhyar, Dane. *Person Centered Astrology.* Santa Fe, NM: Aurora Press, 1976.

———. *The Lunation Cycle.* Santa Fe, NM: Aurora Press, 1967.

Smith, John E., Harry S. Stout, and Kenneth P. Minkema, eds. *A Jonathan Edwards Reader.* New Haven and London: Yale University Press, 1995.

Synan, Vinson. *The Holiness-Pentecostal Tradition: Charismatic Movements in the Twentieth Century.* 2nd ed. Grand Rapids, MI: William B. Eerdmans, 1997.

Tarnas, Richard. *Cosmos and Psyche: Intimations of a New World View.* New York: Viking, 2006.

———. *The Passion of the Western Mind: Understanding the Ideas that have Shaped Our World View.* 1991. Reprint, New York: Ballantine, 1993.

———. *Prometheus the Awakener.* Woodstock, CT: Spring Publications, 1995.

Thomas, Keith. *Religion and the Decline of Magic: Studies in Popular Beliefs in Sixteenth- and Seventeenth-Century England.* 1971. Reprint, London: Penguin Books, 1991.

Weber, Max. *The Protestant Ethic and the "Spirit" of Capitalism and Other Writings.* 1905. Edited, translated, and introduced by Peter Baehr and Gordon C. Wells. New York: Penguin Books, 2002.

———. "Science as Vocation." 1919. In *From Max Weber: Essays on Sociology,* translated and edited by H. H. Gerth and C. Wright Mills. New York: Oxford University Press, 1946.

Winslow, Ola Elizabeth. *Jonathan Edwards 1703–1758: A Biography.* New York: Macmillan Company, 1941.

The Shape of Nihilism

A Cosmological Exegesis of Nietzsche's "The Madman"

Joseph Kearns

> *What were we doing when we unchained this earth from its sun?*
> — "The madman," in *The Gay Science*, section 125.

God is Dead: Nihilism and Cosmology

This essay is an attempt to understand a single passage of Nietzsche's writing—section 125 of *The Gay Science*, "The madman"—by drawing out its cosmological dimension. A deeper understanding of this passage will help to clarify a central theme of Nietzsche's thinking: nihilism, and the means to its transcendence. This theme is represented symbolically by the phrase that forms the axis of the passage: "God is dead." This phrase will be examined with the intention of working towards a more effective grasp of the philosophical origins behind the crisis of values currently encountered in Western culture and beyond. Under the aspect of globalization, the consequences of nihilism pertain not only to the West, nor even solely to the human species. Nihilism has become a planetary phenomenon and its repercussions are proportionate to its pervasiveness. By recognizing the significance of nihilism, specifically in its cosmological dimension, a decisive step might be taken towards its overcoming.

As there are different modes of understanding *cosmology*, it might be helpful to suggest here a provisional definition, specifically as it relates to the phrase "God is dead." A fuller and more adequate understanding of this word will emerge during the course of the essay. Cosmology, as I choose to employ the term, refers to the fundamental *sense of place* engaged by the human in relation to the universe as a whole. This sense of place is based in the primary perceptual cognition and existential awareness of the universe in which we find ourselves, informed by the categories both of spatial and temporal extension. As space-time constitutes the four-dimensional matrix of existence, any given cosmology will be shaped by its spatiotemporal understanding, the sense of the "surround." It is important to recognize that, from a cosmological standpoint, the use of spatiotemporal language is not merely analogical but has a direct correlate in the psychological experience of four-dimensional reality.[1]

This definition considers cosmology's most general function of framing a worldview, making explicit the participatory role of human consciousness in that process.[2] To understand a cosmology, therefore, involves an assessment of its relationship to human thought. Thus, a cosmology—our sense of place in relation to the universe—is an encompassing framework that seems to precede thought and yet proves susceptible to thought's activity. As such, a cosmology is not fixed, but capable of paradigmatic evolution.[3] It shapes and defines a collective reality, yet it may also be perceived on an individual level. Though prevalent and enveloping, a cosmology persists largely as an unconscious background to daily life and only in special instances does it become an object of the conscious mind. As a coherent and integrated vision, a cosmology may be expressed variously in mythopoetic, scientific, or philosophical language. Even when it cannot be articulated coherently, a cosmology forms the a priori backdrop for the unfolding of any particular life story and sets the stage for its ultimate range of possibilities. What we call "the universe" surrounds us and yet is also within us, existing "through" us, as we are "part" of it. Finally, conceived as a whole, a cosmology eludes the specificity of literal definitions and may often be more readily apprehended through metaphors and images, for instance in sacred iconography.[4]

As it is inflected through the human mind, any cosmology (as here defined) inevitably includes a deeply embedded, existential component. Nietzsche was writing from within a post-Copernican cosmology that had displaced the human from the center of an ordered series of encompassing spheres to the indeterminate margins of an infinite expanse. This was bound to impact not only the psychological experience of the spatiotemporal matrix but also the human sense of belonging in the universe.[5] In the passage to be discussed, it will become apparent that when Nietzsche uses the phrase "God is dead," he is referring to something much broader and more multivalent than can be adequately explained in sociological, historical, or even philosophical terms.[6] From the evidence of Nietzsche's own account, the cosmological dimension is fundamental to the task of reaching a satisfactory comprehension of this phrase. To recognize the significance of the nihilism that Nietzsche describes, we must understand the cosmological framing that made it possible.

A Note on Interpretation

The method I adopt in this essay is a qualitative, close reading of a single and relatively short passage of Nietzsche's writing.[7] Given the widespread citation of "The madman," it is surprising to find, at least among the major scholars, a dearth of detailed exegeses that work with the interior meaning of the text itself.[8] Even when considered within its own limits as a self-support-ing piece of philosophical literature, this passage contains a depth and rich-ness that justifies a closer look than that afforded by verbatim reproduction alone. This is neither the first nor the only place that the phrase "God is dead" appears in the Nietzsche corpus.[9] Nevertheless, "The madman" is granted distinction by managing to condense a manifold network of abstruse ideas, elaborated throughout different sections of *The Gay Science*, into a compound unity charged with lyrical poignancy. If "The madman" is a skillful concentration, the phrase "God is dead" is its titrate. This is what makes both the passage and the phrase so quotable.

There are no facts, only interpretations.[10] Voiced by Nietzsche over a cen-tury ago, this insight is now an established premise of postmodern literary deconstruction. Any interpretation is shaped before it even begins by a host of informative factors—psychological, biographical, sociocultural—many of which may remain unconscious to the interpreter. The act of interpretation is selective, both in what it chooses to analyze and in what it chooses to see in the analysis.[11] It is in the hope of making my own perspective more transparent that I offer the following item of personal biography as a prelude to the main text. Recounting this brief narrative has the added advantage of illustrating the existential impact of ideas that may otherwise seem abstract or overly generalized. Furthermore, it exemplifies an underlying proposition of this paper that is especially relevant to "The madman," and perhaps to Nietzsche himself: that the processes operating on a grand scale in human history may also be localized in a personal life story—the phylogeny of a culture recapitulated in the ontogeny of individual experience.

God is Dead: A Biographical Account

The event that I shall relate signifies a turning point in a process of transformation that has covered many years of my life and that continues to exert an influence on my most fundamental sense of belonging and purpose in the cosmos. For the sake of brevity, I describe just one discrete moment of my life story that happened at a particular time and place. However, the larger dynamic remains in the background as the compelling reality that this moment epitomizes. I draw encouragement from Nietzsche's passion for engaging ideas with his total being: "I have at all times written my writings with my whole heart and soul: I do not know what purely intellectual problems are."[12]

I was born into a liberal Catholic family. My father inherited his religion from Irish parents, my mother converted to the Church in her twenties. I was educated in Catholic schools and served as an altar boy for ten years in my local church, where I was initiated in the first four of the seven sacraments: Baptism, Confession, the Eucharist, and Confirmation. For the better part of my childhood, I took the existence of God for granted as a preestablished fact. The God I knew was like a member of the family, a personal God, who had a direct concern and interest in my life and in all individual lives. I *knew* God to be present always and everywhere; it was a knowing that preceded the concept of belief—I could contact Him at any time and feel His presence. For better or worse, my actions were not unnoticed: their moral standing was registered at a higher level and would be accounted for. Hell and the Devil were real but feared less than offending God, and thereby being alienated from His love. I remember a palpable sense of relief after each Confession. When my turn came to die, failing any cardinal offense, I would be taken to Heaven to be reunited with lost family members.

The events of my life, the trials and successes, down to the smallest details, were situated within an all-encompassing context that connected them to the wider community and to the world at large. I was led to understand that God had created the whole universe, over which he presided with a paternal, loving care. Thus the universe was fundamentally meaningful and purposeful, even those elements that were beyond my own limited understanding. All suffering would somehow be subsumed and ultimately vindicated by a greater providence. The universe made sense and love was at its core. There was a sense of *containment*: everything had a place and was held together by a higher principle.

My elementary school and local church were named after Saint Thomas More (1478–1535), the British civil servant and author of *Utopia*, who was executed by Henry VIII on the charge of treason, following his refusal to condone the King's gradual separation from the Roman Catholic Church (precipitated by annulment of the royal marriage with Katherine of Aragon). As a martyr and saint, it seemed a matter of course that Thomas More would have entered Heaven with the minimum of bureaucratic hassle at the gates. He was a shining example of someone who, despite prodigious worldly success and luminous prestige, was prepared to renounce everything, even his own life, for the greater glory of God. His last words on the scaffold were reported to be: "The King's good servant, but God's first." I imagined a close cohabitation and friendship between God and Thomas More in Heaven.

At the age of fourteen, I became involved in a conversation with my brother and two friends about the prospect of death, which inevitably seg-ued into a discussion on Heaven, and specifically the conditions required for entry. Saint Thomas More was mentioned: he of all people, having died for his faith, must surely have been certain of admittance. My brother then asked a disturbing question, which I paraphrase: What if, when Thomas More died, there was *nothing*, just *blackness*? What if, instead of the warm embrace of Heaven, nothing awaited Thomas More but annihilation?

This thought suddenly grabbed me. After the conversation ended, I could not get the thought out of my mind. No matter how hard I tried to dismiss it as a wayward and destructive notion and continue with my usual business, it stuck with me and grew until it became a discomfiting obsession. I had experienced the worldview of my childhood—the ordered sphere created by God, with Heaven above and Hell below—as the primary fact that made all other facts possible; but there was nothing *from within* that worldview itself that I could use to counter my brother's argument. You either believed it or you did not. Until then, any alternative to my worldview had appeared on the *outside*—something that *others* believed. Only after this conversation with my brother did I begin to consider that an alternative worldview might be possible for me. As if by magic, I started to encounter, in news media and books, arguments that supported such an alternative: the real world was comprised solely of physical matter and was ultimately purposeless; death was final and Heaven a myth. I tried to sustain my belief in Heaven, but it began to lose its hold. What if it was all a story that people invented to comfort themselves? Having known this thought for myself, I could not un-know it. The purity of my childhood Heaven had been sullied and I could not remove the stain.

One night this thought—the thought of *Nothingness*—overshadowed all others. As a child, it had been a favorite pastime of mine to lie in the grass for hours and admire the night sky. The star-studded canopy above seemed mysterious and enchanting, its vastness and beauty a cause for wonder and awe. This night, with the thought of Nothingness at the forefront of my mind, I lay down once more to draw familiar comfort from the stars. Comfort was not what I received. Within a few moments, I was terrified. It was as if someone had pulled the plug on the universe, and all meaning, purpose, and love had been drained away. A palpable sense of dread rose up in me before infinite space that suddenly seemed devoid of the loving care I had been accustomed to see in it. God—a childhood fantasy perpetuated by adults who were either irresponsible or thoughtless—was no longer there: He had died. What remained was a stark world governed by natural forces that were oblivious to the anthropomorphic projections of desperate humans. Embroiled in such a savage and impersonal reality, I thought, the most pressing law must be the survival of the fittest. Morality had no greater sanction. Selfishness was, at root, the primary human motive.

Never before or since have I felt so utterly alone in the universe. I appeared to myself as an alien being, strangely conscious of myself and yet disconnected from everyone and everything around me. The thought of Nothingness had interfered with my regular responses to the living world. My desires and aversions would rise and fall continuously without the larger context by which they had been made meaningful. The ultimate sanction of any desire lay in its own time-bound trajectory: it would come, and then it would go, to be replaced by a state of absence and suffering. I felt as if I had woken up from an illusion that still held countless minds captive, but I had woken *into* a nightmare. The way out—death—seemed no less horrifying a prospect. The canopy above no longer sheltered but oppressed me. The stars—distant, cold, and merciless—cared nothing for my fate. How cruel and absurd was life!

In recounting this story, I wish to emphasize the cosmological dimension. The possibility of God's nonexistence had haunted me for some time following the conversation with my brother. But the moment this realization *landed*, as an unavoidable existential realization, was the moment I looked up at the stars. The God I had lost, the God who had died, was not simply the biblical God I had learned about in school, who had spoken to Moses on Sinai and died on the cross in the person of Jesus. The God I had lost was the ever-present and ubiquitous force of cosmic love that underlay the workings of the universe and provided them with order and purpose. *God* had died and the *universe* was drained of meaning. The existential absence was for me experienced *cosmologically*, as a shift in my fundamental relation to the universe as a whole. A simple thought had entered my consciousness and the world was not the same. The psychological experience was of a shift in the world *around* me, like a change in the climate, and yet it seemed to penetrate to the very core of my being. The thought had occurred "within" me, in my mind, and yet the universe itself seemed colder and starker, the Earth a random rock in fathomless space. The sense of *containment* had evaporated as the universe spilled out in all directions.

One more detail is worth acknowledging, apropos of this personal story. Following the conversation with my brother, a tremendous tension had accumulated within me between the womb-like worldview of my childhood and an emerging worldview that considered the universe as essentially meaningless. On the night the tension broke, I remember in a sense *surrendering* to the new cosmology—the thought of Nothingness. Something in me could no longer resist it, and painful though it was, I made a conscious decision to value what I *saw* as the truth over and above what I *wanted* to be true. In brief, the dissolution of the Catholic worldview of my childhood involved a degree of participation from me—it was not entirely outside of my own volition. The event—the death of my God—did not just happen *to* me; it occurred with my own collusion. This aspect of the event—the participatory role of human consciousness—is fundamental to the analysis that follows.

Exegesis

The Kaufmann translation of section 125 of *The Gay Science* is here reproduced in full. Letters in superscript indicate the subtitles of the analysis below.

The madman.[A]— Have you not heard of that madman who lit a lantern in the bright morning hours,[B] ran to the market place, and cried incessantly: "I seek God! I seek God!" —As many of those who did not believe in God were standing around just then, he provoked much laughter. Has he got lost? asked one. Did he lose his way like a child? asked another. Or is he hiding? Is he afraid of us? Has he gone on a voyage? emigrated? —Thus they yelled and laughed.

The madman jumped into their midst and pierced them with his eyes. "Whither is God?" he cried; "I will tell you. *We have killed him*[C]—you and I. All of us are his murderers. But how did we do this? How could we drink up the sea? Who gave us the sponge to wipe away the entire horizon? What were we doing when we unchained this earth from its sun?[D] Whither is it moving now? Whither are we moving? Away from all suns? Are we not plunging continually? Backward, sideward, forward, in all directions? Is there still any up or down? Are we not straying as through an infinite nothing? Do we not feel the breath of empty space? Has it not become colder? Is not night continually closing in on us? Do we not need to light lanterns in the morning? Do we hear nothing as yet of the noise of the gravediggers who are burying God? Do we smell nothing as yet of the divine decomposition? Gods, too, decompose. God is dead.[E] God remains dead. And we have killed him.

"How shall we comfort ourselves, the murderers of all murderers? What was holiest and mightiest of all that the world has yet owned has bled to death under our knives: who will wipe this blood off us? What water is there for us to clean ourselves? What festivals of atonement, what sacred games shall we have to invent? Is not the greatness of this deed too great for us? Must we ourselves not become gods simply to appear worthy of it? There has never been a greater deed; and whoever is born after us—for the sake of this deed he will belong to a higher history than all history hitherto."

Here the madman fell silent and looked again at his listeners; and they, too, were silent and stared at him in astonishment. At last he threw his lantern on the ground, and it broke into pieces and went out. "I have come too early," he said then; "my time is not yet. This tremendous event is still on its way, still wandering; it has not yet reached the ears of men. Lightning and thunder require time; the light of the stars requires time; deeds, though done, still require time to be seen and heard. This deed is still more distant from them than the most distant stars—*and yet they have done it themselves.*"

It has been related further that on the same day the madman forced his way into several churches and there struck up his *requiem aeternam deo.* Led out and called to account, he is said always to have replied nothing but: "What after all are these churches now if they are not the tombs and sepulchers of God?"

A. *The madman*

For the majority of *The Gay Science*, Nietzsche addresses the reader directly. Indeed, one of the marks of his overall style is the prevalence of the vocative: Nietzsche addresses *you*. Throughout Nietzsche's work, one consistently encounters imperatives and rhetorical questions. The transference of the speaker to a third person is a deliberate literary device that is all the more distinctive for its infrequency in Nietzsche's writings. The most significant usage of this device is with the character who appears at the end of Book Four of *The Gay Science* (section 342) and who dominates the eponymous book that followed: Zarathustra.[13] What the use of the third person achieves in *Thus Spoke Zarathustra* applies to "The madman" also: it allows for a shift in tone that elevates poetic insight above the philosophical skepticism that had kept Nietzsche, until then, from attributing metaphysical status to statements about human psychology. The passionate emotional investment, that elsewhere lends forcefulness to Nietzsche's argument and questioning, is given an immediate voice through another. As with *Zarathustra*, the prophetic language of "The madman" is facilitated by a third person narrative that grants its author an artistic license loosened from the familiar restraints of self-conscious irony.[14]

The use of the third person, therefore, enacts a sustained ambiguity: Is Nietzsche reporting the actions of a madman or is he speaking for himself? More tellingly, is this the private opinion of one individual—the ravings of a lunatic—or does it possess a wider relevance? What exactly is "mad" about this madman? Notice that the madman is not introduced as someone whose sanity is in question, owing to some incongruous and disruptive behavior. The diagnosis has taken place before he even arrives on the scene: there is no "man" apart from the "mad." Whatever the madness of the madman entails, it is integral to his presented identity.

Today, "madness" is a controversial and somewhat outdated term. Advances in twentieth-century psychiatry, psychology, and related philosophical theory have exposed some of the ethical shortfalls inherent to the act of categorizing and stigmatizing certain forms of behavior according to societal and political expedience.[15] Madness may only be defined by implicit reference to its opposite, "sanity," both of which exist on the scale of human potential, the latter being the more tightly circumscribed and socially validated. It is fitting, therefore, that the madman appears in direct contrast to the people "standing around in the market place." Evidently his predefined madness owes something to his appearance and behavior, which are strange only by comparison to those he addresses. But what difference does this appearance indicate? At the outset, a rift is established between two ways of being, one represented by the madman, the other by the market goers. The dramatic tension of the piece rests on this meeting between different worlds of consciousness. The ambiguity remains: Does the madman bring something of worth from his world or is it nonsense? Is he mentally incoherent or does he embody an uncomfortable truth?

The overall tenor of the piece, the cogency and coherence of the madman's words, and the astonishment of the listeners in the penultimate paragraph, would suggest that the epithet "mad" in this instance is more indicative of a collective short-sightedness than an individual's psychosis. This man has something important to say. In his role as madman he expresses a truth that the surrounding culture either does not see or chooses to ignore. In this regard, his madness is closer to Hamlet's than King Lear's.[16]

But it will not suffice to explain away the madman as either a prophetic truth-teller or a sociological anomaly. Why did Nietzsche choose to title this piece "The madman," rather than, for instance, "God is dead" or "The death of God"? At this point, it will help to employ a concept from the field of existential psychology developed by R. D. Laing: *ontological insecurity*.[17] As Laing describes, many people assume as primary facts certain qualities and conditions that constitute their sense of selfhood. These include: a sense of self coextensive with the physical human body, an experience of being a real entity that has continuity in space-time, and a belief in the basic reality of the world and others as distinct from oneself. The existence of an integrated and cohesive ontological security depends upon a presupposed fundamental distinction between "self" and "world." Such a distinction makes possible functional interactive relations between different beings.

As Laing asserts, it is possible that these "primary facts" may not be experienced as such. Thus, a person may not have an enduring sense of the self either as being coextensive with the physical human body or as possessing continuous and stable existence in space-time. A foundational assumption of the basic "reality" of the world and others may not be present. The "ontologically insecure" person may hold the premises that make ordinary life possible (and even desirable) to be unanswered questions. For the person who lacks this basic ontological security, events in daily life that may seem trivial to others may quickly become problematic and overwhelming.

As suggested above, Nietzsche's madman seems to present as a total and integrated personality: his argument is pointed and coherent even as it is abstruse and poetic; his posture seems designed to make a certain impact; and he has enough self-assurance to believe in himself and his message. On the evidence of the passage, the madman seems to possess a primary ontological security in regard to his own person. But the same may not be said of his *relation to the world*. According to the madman, the premises that previously constituted the reality of the universe—its continuity and coherence in space-time, its basic ontological framing and ground, its supposed moral purposes—have become unanswered questions. The effect is *disorientation*. To the extent that his ontological security depends upon the coherence and meaningfulness of the world picture that includes him, the madman is mad.[18]

If we apply the concept of ontological insecurity to the self's relation to the world, it becomes possible to speak of *cosmological* insecurity. The foundational sense of place engaged by the human in relation to the universe has become ontologically insecure. It is the contention of the madman that this "cosmological insecurity" is not peculiar to one individual. The shift he describes is one in which everyone is implicated, including those buying and selling in the market place. The message contains a psychological validity for both the individual and the collective. However, this madman is unique in that he dares to embody the psychological consequences and responsibilities of this basic cosmological insecurity. The others in the market place seem content to carry on as usual, as if nothing had happened.

Clearly Nietzsche's choice of title is not without reason. The *cosmological* event signified by the phrase "God is dead" has huge *psychological* implications. The madman is a mouthpiece for a cosmic madness. And so, out of the death and decomposition of one god, the latent presence of a different divinity may be felt, as it could in so many ways throughout Nietzsche's life and work: Dionysus, the god of creation through destruction, the god of cosmic madness.[19] It is in this spirit that Zarathustra says, "Where is this madness, with which you should be cleansed?"[20]

B. *Who lit a lantern in the bright morning hours*

The first sign of the madman's erratic tendencies occurs at the beginning of the passage: he lights a lantern *in the bright morning hours*. Besides contributing to the strangeness of the madman's appearance, the act of lighting a lantern in broad daylight conveys a more general sense of incongruity. Considering the purpose for which they were invented, it is wasteful and pointless, some would say mad, to use lanterns when their light is not needed. Here a type of human behavior is displayed at odds with the natural order. Something is awry.[21]

The overt symbology of this act also shows the madman to be something of a performance artist. In the second paragraph he asks, "Do we not need to light lanterns in the morning?" Then, in the silence that follows his main speech, he breaks the lantern on the ground, extinguishing its light, and says, "I have come too early . . . my time is not yet." The act of lighting the lantern is precocious—it is meant for the night. The madman seems to be suggesting that a shift has happened in the surrounding climate—expressed in cosmological terms—that will require the lighting of lanterns. "Has it not become colder?" he asks, "Is not night continually closing in on us?"

A lantern's flame is sparked by a human agent; it burns with a fuel supplied by its steward. The time for using lanterns is when the light of the sun is no longer sufficient. In section 343 of *The Gay Science*, Nietzsche employs the imagery of sunset in reference to the statement "God is dead": "Some sun seems to have set and some ancient and profound trust has been turned into doubt . . . our old world must appear daily more like evening."[22] Later in the same paragraph of section 343, Nietzsche refers to this as "an eclipse of the sun whose like has probably never yet occurred on earth." In *Thus Spoke Zarathustra*, a similar motif recurs frequently: the "great noontide," when the sun has reached its highest point and must begin to "go down."[23] In each case, the symbol balances the inexorable autonomy that is proper to celestial bodies with an encouraged human response: a conscious detachment from reliance on the luminosity and warmth of an objective divinity, and the consequent need to initiate, from a human starting-point, the creation of meaning and values—in short, to live in our own light. This is the task for which Nietzsche envisioned the *Übermensch*, as presaged in the madman's words: "Must we ourselves not become gods simply to appear worthy of it?" This transition, from the sunlight of God to the lanterns of the *Übermensch*, is emphatically affirmed in *Thus Spoke Zarathustra*: "'*All gods are dead: now we want the Superman to live*'—let this be our last will one day at the great noontide!"[24]

C. We have killed him.

As Karl Jaspers has remarked, Nietzsche does not have his protagonist declare, "There is no God," nor does he say, "I do not believe in God."[25] In fact, the madman may be distinguished from his interlocutors in that he self-avowedly *seeks* God. In contrast, it is reported that many of the bystanders "did not believe in God." Far from discerning any cause for alarm, they mock the notion of *seeking* God with a series of sarcastic comments. For them, the joke rests on the absurdity of the idea that God himself could disappear. The burden of choice exists entirely *on their side* as a question of belief or unbelief. Reduced to an issue of philosophical, theological, or even psychological preferences, the existence or nonexistence of God is of no overriding concern, and the daily business of the market place may continue unimpeded.

It is therefore left to the madman to perceive an *absence*, which he attributes to a source that obtains beyond his personal attitude. The madman brings news of an *event* that has already happened. As mentioned above, this event has a measure of autonomy, like the sun setting. It manifests as a change in the "surround," like a change in the climate, making the air colder and the night closer. The metaphor of death emphasizes the autonomy and the finality. There is no talk of resuscitation: God has begun to decompose—a process as natural as death itself. In the penultimate paragraph, having broken his lantern, the madman asserts the autonomy of God's death apropos the widespread unconsciousness concerning it: "This tremendous event is still on its way, still wandering; it has not yet reached the ears of men." As Nietzsche insists in section 343 of *The Gay Science*, the death of God may be described as an "event," initiated at a level beyond general apprehension, yet destined to have immanent and widespread psychological consequences. In addition, a line from Nietzsche's notebooks, in reference to the phenomenon of nihilism, accentuates the spontaneous aspect of the process: "What does nihilism mean? *That the highest values devaluate themselves.*"[26]

But the failure of the people in the market place to comprehend the autonomy of this event does not bring them closer to accepting any sort of responsibility. According to the madman, not only do they miss the significance of God's death, they also do not adequately recognize the extent to which they themselves are profoundly implicated. For as the madman stresses repeatedly, though this event has happened before it has even been realized, it paradoxically includes a large measure of human participation: "This deed is still more distant from them than the most distant stars—*and yet they have done it themselves*." This involvement in the death of God extends beyond a mere reception of the consequences: humanity is actively responsible. This "great deed" has happened not just *to* but *through* the human psyche: "*We have killed him*—you and I. All of us are his murderers." The madman urges that the death of God be encountered as an unavoidable, present reality, so that its implications may be squarely faced. To this end, he demands that the human role in this state of affairs be fully acknowledged.[27]

In its essentially paradoxical nature—as a deed that is at once as distant as the stars and yet pressingly close, coming towards us and yet proceeding from us—the death of God, as described by Nietzsche, conforms to the definitions of cosmology that were outlined at the beginning of this essay. As a cosmological dynamic, the death of God precedes thought, and yet is somehow susceptible to thought's activity; it describes a kind of paradigmatic evolution in the nature of the "surround"; as part of the collective unconscious background to daily life, it possesses a deliberate autonomy that to some extent remains mysteriously responsive to human choice and interaction. Even though shaping and defining a collective reality, it may be perceived and experienced existentially on an individual level, whether that person is a madman or a fourteen-year-old boy.[28]

D. What were we doing when we unchained this earth from its sun?

But how is it possible to *kill* God? "How could we do this?" asks the madman. Having been told of the culprits, one might then expect a description of the weapon, the method by which it was administered, and the possible motives for its use. Instead, through a series of rhetorical questions, the madman begins to indicate what this murder actually involves. "How could we drink up the sea?" he asks, "Who gave us the sponge to wipe away the entire horizon?"

The sea and the horizon are common motifs in Nietzsche's writings, often occurring together.[29] As used by Nietzsche, the image of the sea often evokes the groundlessness and dangers of being adrift, unmoored from a traditional ontology, as well as the radical freedom and openness to new horizons. To "live dangerously" is necessary to the spirit of adventure.[30] But the sea also implies another meaning: the matrix of the spatiotemporal "surround," all-encompassing like the ocean, extending as far as the eye can see. The madman delivers a shock that would not be out of place in the Pentateuch: it is as if the sea itself has disappeared. The prophetic reach of the statement makes it more than a personal confession or philosophical conceit. A tide is turning.[31] The death of God means a fundamental change in the existential fabric of the surrounding world. One may recall the analogy from my own biographical account: it was as if the plug had been pulled on the universe.

The horizon metaphor supports the overall effect. As a spatial term, a horizon circumscribes a total area. It defines the background against which events in the foreground unfold. It is the a priori shaping of the landscape and as such it predicates a range of movement. Any given trajectory takes place within, and moves towards, a horizon. It is the boundary between the known and the unknown. What the madman describes is a loss of the entire horizon. The phrase "God is dead" refers to a change in the *background*—the cosmological stage upon which the human drama is enacted. But how is this so? How did God, when alive, provide such a horizon?

As the surrounding passages in *The Gay Science* make clear, the God who dies is not merely the Christian God of the Bible and the Church, but the God who pervades the entire cosmological sphere.[32] On the one hand, in His *immanence* as a cosmic reality, God ensured the intrinsic order and ultimate purpose of the world. As an expression of the master plan of God, the universe gained a linear trajectory and spatiotemporal continuity and cohesion as a whole. This God imbued the world with His presence both spatially (in the "surround") and temporally (from the beginning to the end of time). On the other hand, in His *transcendence* as a metaphysical being, God granted the world of space-time its external framing. Suffering, injustice, impermanence, and death could be vindicated and made meaningful by reference to *another world*. Whether situated at the furthest reaches of the stars, or placed in a realm beyond space and time, the God of Heaven represented the phenomenal world's ultimate horizon, its "container."

It is not hard to see why this God flourished in a geocentric universe. Heaven and Hell correlated with the Above and the Below. The Earth was centrally situated in a harmonic order of encompassing spheres, which proceeded outwards from the terrestrial core in ever-widening circles. The human being was assigned an integral place of belonging. A single human life had an aim, a trajectory within God's broader purpose. Death, as life's horizon, was not a termination but led to greater things. With these considerations in mind, and with the benefit of hindsight, the cosmological aspect of the death of God seems inevitable. The Copernican shift, as it dislodged the human on Earth from the cosmological center, also distanced and displaced the heavens. Even God, therefore, seems to have been affected by the Copernican revolution; as the madman might say, perhaps He is its greatest casualty.

Nietzsche emphasizes the spatiotemporal disruption in the cosmology, and the psychological upheaval that it entails, in the following passage:

> What were we doing when we unchained this earth from its sun? Whither is it moving now? Whither are we moving? Away from all suns? Are we not plunging continually? Backward, sideward, forward, in all directions? Is there still any up or down? Are we not straying as through an infinite nothing? Do we not feel the breath of empty space?

These lines describe the psychological experience of a radically altered cosmological reality: there is an overwhelming sense of the chaos and absurdity of existence, of the purposeless drifting in the void, disconnection, cosmic alienation, and profound disorientation at the lack of overall picture and metaphysical horizon.[33] The effects of this spatiotemporal entropy have also been described by John Berger, who writes the following in regard to the right-hand panel of Hieronymus Bosch's *The Garden of Earthly Delights*, painted at the beginning of the sixteenth century:

> There is no horizon there. There is no continuity between actions, there are no pauses, no paths, no pattern, no past and no future. There is only the clamour of the disparate, fragmentary present. Everywhere there are surprises and sensations, yet nowhere is there any outcome. Nothing flows through: everything interrupts. There is a kind of spatial delirium.[34]

Figure 1 **The Garden of Earthly Delights, Hieronymous Bosch (right panel)**

E. *God is dead*

A cynic may remark that there is actually nothing new about the death of God. From the perspective of the Christian tradition, it is held that God died on a cross two thousand years ago. This "first" death, through the figure of Christ, marked the beginning of an era, expressed temporally by the establishment of an axis point in linear time (BC/AD) and in subsequent calendrical reforms. The God to whom Nietzsche refers is unmistakably the God of Christianity.[35] In "The madman" it is possible to see references to the Passion and Resurrection of Christ.[36] In the third paragraph, the "murderers of all murderers," the blood-guilt and washing of hands, evoke the trial of Jesus before Pilate; the "festivals of atonement" recall the Christian rituals performed in memory of this event. In the final paragraph, the "tombs and sepulchers of God" and the clever twist of words taken from the Mass for the dead—from *requiem aeternam dona eis, Domine* (eternal rest give unto them, O Lord) to *requiem aeternam deo* (eternal rest unto god)—further elaborate the connection to Christian rites.

But the "second" death of God announced by Nietzsche is appointed to mark the end of the age that began with the first. As indicated above, the dying of a god entails more than a simple exchange of religious preferences, an affirmation of belief or disbelief; it means the evaporation of an entire "containing" mythos. On the evidence of Nietzsche's own account in "The madman," the death of God is a compact symbolic formulation for a complex of causes and conditions that together constitute a sweeping cosmological dynamic. This dynamic is effectively a revolution in worldviews that takes place coactively at the interface between the human mind and the universe itself. This amounts to a total vision of the world and the human place within it, expressed concretely through the experience of the enveloping matrix of space-time.

It is the cosmological dimensions of the death of God that make it so all-pervasive and epoch-shifting. First, as I outlined in my biographical account, this God is an emphatically *personal* God, present always and everywhere, caring for each person individually, watching and judging, the principle of a permeating cosmic love that operates by means of an inscrutable intelligence. Stripped of this surrounding presence, the world may appear a stark and savage place where the rule of "might means right," the survival of the fittest, exists as a primary truth. Nietzsche's attempt to find the origin of moral tenets and metaphysical speculations in biological, amoral impulses—and his subsequent discovery of power as *the* primary motive force—follow directly from the absence of a loving and personal God. Love, divested of any metaphysical properties it may once have had, is reduced to a feeling-state with a provenance in the will to power.

Second, the death of God deprives the universe of its *telos*. The Christian mythos provided an account of the origins and destination of the universe in terms of an ultimate purpose. The human had a central role to play in that story. Where Copernicus and Darwin had relegated the human to a miniscule bit-part in the cosmological story, the death of God disposes of the story itself.[37] As the story is lost, so is the aim. Consciousness becomes an accidental epiphenomenon in a random cosmic flux. Without a discernible reason for being—no why, no whence, no whither—both human self-reflexive consciousness and the surrounding universe appear as fundamentally absurd.

Third, as Heidegger asserts, the death of God refers to the whole of the suprasensory world in general and therefore comes to stand for the end of metaphysics.[38] As the metaphysical "horizon" disappears, an ontological entropy ensues in which the unitive order and coherence of the phenomenal realm are destroyed. Properly speaking, there is no "universe" as a whole because there is no greater reality to define it against. As Nietzsche says, there is no "world" as such, with a perceptible beginning and end.[39] The death of God ends the singular determination of a monotheistic cosmos.[40] The ontological fragmentation corresponds with the spatial disorientation initiated by the Copernican shift. The series of ordered spheres, with the earth at the center, is replaced by a shapeless fragmentation with no discernible center. All is chaos.[41]

Fourth, and following from these last two points, God's absence means an absence of *truth*. The revolution in epistemology advanced by Kant, by emphasizing the innate limits to the possibilities of human thought, rendered any knowledge about the world itself radically provisional. If no metaphysical conclusions could be drawn from empirical information, nothing definitive could be asserted about the nature of reality itself, let alone the reality of a transcendent God. As developments in biblical exegeses had demonstrated, even the Bible was not immune to textual scrutiny, and proved fallible to the test of truth when subjected to interpretive techniques that disclosed internal inconsistencies.[42] According to Nietzsche, the spirit of "truthfulness," that takes truth as a primary value, begins to undo the Christian moral structures that fostered it through the ages.[43] God may no longer be cited as the ultimate arbiter of truth. Even science, despite its spirit of intellectual integrity and methodological rigor, is based on ultimately unprovable premises that inhibit the truth-value of its findings.[44] The rise of scientific "truth" in the modern age assists in the "killing" of God; but the death of God means the end of any a priori standard of truth. This leads to Nietzsche's "ultimate skepsis."[45]

Fifth, with the death of God, morality is deprived of its most immediate ontological ground, and thereby weakened when considered as a cosmological force. The universe comes to be seen as devoid of any inherent and absolute moral structures. Morality is arbitrary and relative, and valuation is exposed as a human activity, perhaps *the* fundamental human activity.[46] According to Nietzsche, morality is not only a question, but a problem.[47] As with love, it may be traced to unconscious and essentially amoral impulses in the human organism. Nietzsche's endeavor to expose the amoral foundations of morality, its cosmological "nakedness," as a prerequisite to the deliberate creation of new values, proved to be an overarching theme of his life's work.[48] Without a sovereign principle of goodness in the person of God, goodness itself is relativized and must seek new justifications.

Finally, the death of God occurs in tandem with the destruction of the concept of Heaven, and so with the ready-made solution to human biological death. If God is not there waiting for the human soul after death, then what is? As my brother had asked so persuasively, what if nothing awaited Thomas More but the annihilation of his consciousness? The death of God emphasizes the finality of the death of the individual. In turn, this strengthens the overall sense of the absurdity of a temporal self-consciousness, borne by the short life of the body, enclosed at both ends by an infinity of nothingness. After seeing the death of the tightrope walker, Zarathustra says, "Uncanny is human existence and still without meaning: a buffoon can be fatal to it."[49]

The disenchantment of the cosmos, the loss of aim and story, the end of metaphysics, the disruption of the spatiotemporal framing, the rise of scientific "truth," and the demise of the ontological status of morality—all were part of the same cosmological process by which God perished. As expressed by Nietzsche in "The madman," the death of God is a participatory cosmological event.

God is Dead: The End—and a Beginning

In its expression of cosmological insecurity, or cosmic madness; with its imagery of lanterns in the daylight; as it articulates the paradox of an event that is both individual and collective, transpersonal and personal, autonomous and occasioned through human participation; and as it frames the inevitable psychological disorientation in terms of disruption in the spatiotemporal "surround," with specific reference to the post-Copernican reality—"The madman" describes the cosmological basis for the nihilism that Nietzsche anticipated in Western culture.

Today the power of the phrase "God is dead" derives less from its shocking polemic and more from its widely recognized accuracy in concisely articulating a problematic truth of the modern world—and in recognizing the significance of that truth. As with all death, the death of God is weighted with finality, referring to an ending, a closure, discharged with the force of necessity. But this "divine death" signals the ending of a particular, fixed cosmology—a shift in the fundamental sense of place engaged by the human in relation to the universe as a whole. The phrase "God is dead" is a compact symbolic formulation for a multivalent cosmological complex that transcends statements of assent or refutation—because *it has already happened* as a participatory event. Whatever *belief* one holds about the nature of the world occurs within the broader cosmological *setting* symbolized by the death of God. Those who do not understand, or choose to ignore, the post-Copernican reality still must live in it, and yet it is a worldview engendered through the participation of the human mind.

While Nietzsche articulates how this revolution in cosmology can be terrifying and profoundly disorientating, he was also deeply sensitive to its liberating and empowering dimensions, as well as its capacity to invoke a renewed cosmic curiosity. In his life and writing, Nietzsche did not stop at the recognition of nihilism and its psychological effects but pushed further towards its overcoming through a "revaluation of all values."[50] Nietzsche's madman declares, vis-à-vis the death of God, "There has never been a greater deed; and whoever is born after us—for the sake of this deed he will belong to a higher history than all history hitherto." This affirmation points beyond the initial terror of nihilism to an active sense of responsibility for the invention and discovery of new values, as conscious participants in the creation of our own history. From this perspective, the nihilistic void offers a radical freedom replete with creative potentials. In section 343 of *The Gay Science*, entitled "The meaning of our cheerfulness," Nietzsche writes of this more open and invigorated stance towards the universe:

> Indeed, we philosophers and "free spirits" feel, when we hear the news that "the old god is dead," as if a new dawn shone on us; our heart overflows with gratitude, amazement, premonitions, expectation. At long last the horizon appears free to us again, even if it should not be bright; at long last our ships may venture out again, venture out to face any danger; all the daring of the lover of knowledge is permitted again; the sea, *our* sea, lies open again; perhaps there has never yet been such an "open sea."[51]

Cosmologically, therefore, nihilism may be understood not only as the ending of a fixed worldview but as the "open sea" out of which new vistas of meaning may emerge; that is, nihilism may be considered as one step in an even larger process of paradigmatic evolution, a clearing of the cosmological stage for hitherto unforeseen horizons to appear. As the onset of nihilism is a participatory cosmological event, so may be the means to nihilism's integration and transcendence. And just as Nietzsche describes the cosmological "shape of nihilism," he also evokes the adventure of describing the cosmologies emerging out of nihilism's darkness. The evolution beyond the nihilistic universe at an individual or collective level suggests that God's death is not only an end but has opened the cosmological space for another beginning.

Notes

1. In his article on the ubiquity of spatiotemporal metaphors in the field of transpersonal psychology, Sean Kelly stresses the importance of recognizing both the identity and difference between the thing being described (what it *is*) and the analogues used

to describe it (what it is *like*) to avoid "literalizing the analogy." Spatiotemporal metaphors, or what Kelly calls spatial and temporal "image schemata" (a term proposed by Mark Johnson), are consistently used to indicate transpersonal phenomena that tend to defy literal definition. Kelly argues that four-dimensional existence in space-time (the "positioning" at the intersection of the horizontal and vertical planes) provides the basis for such root analogies, which are so common and foundational that they often pass unnoticed. My argument differs from Kelly's in that the use of spatiotemporal language to describe a cosmology refers directly to the embodied, psychological experience of space-time, the literal sense of the "surround," and so goes beyond "analogy." See Sean Kelly, "Space, Time and Spirit: The Analogical Imagination and the Evolution of Transpersonal Theory," *Journal of Transpersonal Psychology* 34, no. 2 (2002): 73–99. It is also worthwhile keeping in mind the insight developed by Hillman (and alluded to by Kelly), that all theory is, in a sense, metaphorical (or has an analogical component) and "literalism" itself is a pervasive metaphor. See James Hillman, *Re-Visioning Psychology* (1975; repr., New York: HarperPerennial 1992) 149–150.

2. Thus the cosmology described here may be distinguished from its other academic applications either as the scientific and mathematical study of the physical universe or as a branch of metaphysics. For an overview of the distinction, see Brian Swimme, *The Universe Story* (New York: HarperCollins, 1992), 22-24; and *The Hidden Heart of the Cosmos* (New York: Orbis, 1996), 31-32.

3. Owing to its vast scale by human terms, the universe, considered as a singular totality, must be engaged as a hypothesis—the grand whole "pictured" in the mind—into which empirical data and mathematical equations concerning it are subsumed, acting as the scientific support. To infer a general law from empirical data requires a certain "faith" in ideas that cannot themselves be proved empirically and so must be sustained as mental abstracts. In Kuhnian terms, we might say that any overall conception of the universe is a "paradigm." See Thomas S. Kuhn, *The Structure of Scientific Revolutions* (1962; repr., Chicago: University of Chicago Press, 1996), especially chapters 5 and 10. For Nietzsche's perspective on the dangers of drawing premature conclusions concerning the nature of the cosmos as a whole, see Friedrich Nietzsche, *The Gay Science,* trans. Walter Kaufmann (New York: Vintage 1974), 167–177 (sections 109, 110, 112, and 121).

4. With these criteria in mind, it becomes possible to speak of the *archetypal* nature of any given cosmology. Richard Tarnas summarizes the evolution of the archetypal perspective in *Cosmos and Psyche: Intimations of a New World View* (New York: Viking, 2006), 80–85. Jung makes the distinction that archetypes are contents of the collective unconscious in contrast to the feeling-toned complexes of the personal unconscious: see "Archetypes of the Collective Unconscious" in *The Archetypes and the Collective Unconscious, The Collected Works of C. G. Jung,* vol. 9, part 1, trans. R. F. C. Hull, ed. H. Read et al. (Princeton, NJ: Princeton University Press, 1969). Though existing at a collective level, the archetypes may be inflected at the level of an individual, through whom they appear in a particular manifestation. As Tarnas notes, in later life Jung became sympathetic to the view that archetypes possess an autonomy that prevails at a level beyond their mental representations (*Cosmos and Psyche,* 57–59). Hillman also stresses the collective nature of the archetypes, their potent ability to dominate a person's consciousness even while remaining unconscious to that person, and the nebulous quality that makes them liable to be characterized by metaphor (*Re-Visioning Psychology,* xix–xx).

5. The Copernican revolution in astronomy refers to the transition from a geocentric to a heliocentric cosmology initiated by Copernicus in the sixteenth century. Richard Tarnas outlines the psychological and philosophical consequences of the Copernican shift in "The Post-Copernican Double Bind" in *The Passion of the Western Mind: Understanding the Ideas That Have Shaped Our World View* (1991; repr., New York: Ballantine 1993), 416–422. The chief insight of this paper—that the phrase "God is dead" may be interpreted by reference to its cosmological and specifically post-Copernican dimension—owes its inception to this work of Tarnas.

6. The words *secularization* and *nihilism* represent, in sociological and philosophical terms respectively, aspects of the same enveloping reality that the phrase "God is dead" aims to give symbolic resonance. Even when considered to describe the same process, these terms bring out different nuances: "God is dead" expresses through negation the absence of what went before, with the accompanying psychological implications of loss and grief; "nihilism" holds the possibility of an affirmative philosophical stance that has incorporated the preceding absence. "Secularization," of course, refers to social trends that can be measured, without necessarily identifying an underlying cause.

7. The translation is provided by Walter Kaufmann, in Nietzsche, *Gay Science,* 181–182. Where necessary, I have compared Hollingdale's treatment of the same passage, in R. J. Hollingdale, *Nietzsche* (London: Routledge & Kegan Paul, 1973), 65–67. Wherever I quote Nietzsche directly, I have used Kaufmann's translations of the relevant editions, except in the case of *Thus Spoke Zarathustra*, for which I have used the Hollingdale version (1969; repr., London: Penguin, 2003). For a textual exegesis such as the one I am offering, the absence of the original German text is a serious misgiving, one that must remain until I attain a sufficient mastery of the German language. However, I hope to show that working with the English translations, though limiting in certain respects, is no barrier to the basic proposal of my argument—the elucidation of "The madman" by reference to its cosmological dimension.

8. Here are just a few of the places in the Nietzsche literature where "The madman" is quoted either in full or with some marked ellipses: Walter Kaufmann, *Nietzsche: Philosopher, Psychologist, Antichrist,* 4th ed. (Princeton, NJ: Princeton University Press, 1974), 96–97; R. J. Hollingdale, *Nietzsche: The Man and His Philosophy* (Louisiana: Louisiana State University Press, 1965), 167–168; also in Hollingdale's *Nietzsche* (1973), 65–67, and in his introduction to *Thus Spoke Zarathustra*, 14–15; and Martin Heidegger, "The Word of Nietzsche: 'God is Dead'," in *The Question Concerning Technology and Other Essays*, trans. William Lovitt (New York: Harper & Row, 1977), 59–60. An excerpt from the passage is also cited by Tarnas both in *Passion of the Western Mind*, 412, and *Cosmos and Psyche*, 345. Over time, the appearance of "The madman" in diverse contexts has lent the passage a certain emblematic status.

9. According to Kaufmann, the first appearance of the phrase occurs at section 108 of *The Gay Science*. Kaufmann lists subsequent instances, including section 343 of *The Gay Science* and a number of places in *Thus Spoke Zarathustra*. See Kaufmann in *Gay Science*, 167n.

10. See Friedrich Nietzsche, *The Will to Power,* trans. Walter Kaufmann and R. J. Hollingdale, ed. Walter Kaufmann (New York: Random House, 1967), 267 and 327. In this and other publications, both Kaufmann and Hollingdale stress the problems and limitations of using *The Will to Power* as a source text. Once these limitations are realized, its advantages may be judiciously employed. See Kaufmann's introduction to *Will to Power*, xiii–xxiii, and his *Nietzsche: Philosopher, Psychologist, Antichrist*,

6–9; and also Hollingdale, *Nietzsche: The Man and His Philosophy*, 260–72, and *Nietzsche*, xi–xii.

11. One of the outstanding merits of Kaufmann's scholarship was to expose the tendency, especially prevalent among state-sponsored Nazi academics, to co-opt and emend passages of Nietzsche's work with an unscrupulous disregard for their original context, in order to support premeditated ideologies. As Kaufmann describes, the withholding of *Ecce Homo* by Nietzsche's sister enabled the perpetuation of a legend that twisted her brother's views beyond recognition. See Kaufmann, *Nietzsche: Philosopher, Psychologist, Antichrist*, 3–18, 94–95, and 289–292; and Kaufmann's footnote to "The madman" in *Gay Science*, 182. Despite Nietzsche's own forewarnings to later readers, his proclivity for the pithy and controversial turn of phrase made his writings vulnerable to such abuse, a point made by Peter Gay in the introduction to *Basic Writings of Nietzsche*, trans. and ed. Walter Kaufmann (New York: Modern Library, 2000), ix–x.

12. Compare: "One must want to experience the great problems with one's body and one's soul." Both quoted by Hollingdale in his introduction to *Thus Spoke Zarathustra*, 12.

13. Section 342 concluded the first edition of *The Gay Science*, published in 1882. With minor modifications the same section began the prologue of *Thus Spoke Zarathustra*, which appeared in 1883. Within a year, Zarathustra had graduated from bit-part player to lead role. See Kaufmann's commentary in *Gay Science*, vi and 274n. A biographical background to Nietzsche's writing during this time is provided by Hollingdale in *Nietzsche*, 69–74, and in his introduction to *Thus Spoke Zarathustra*, 20–22.

14. It may be argued that using Zarathustra as a mouthpiece enabled the delivery of numerous positive conceptions that would prove decisive for Nietzsche in the project of surmounting nihilism: the *Übermensch*, self-overcoming, the will to power, *amor fati*, and the eternal recurrence. Nietzsche himself declared that he adopted the name Zarathustra as a deliberate antithesis to the original Persian prophet, whose achievement was to grant the moral struggle of opposites an ontological status (see section 3 of "Why I Am a Destiny," in *Ecce Homo*, in *Basic Writings of Nietzsche*, 783). But as Jung has indicated, this ancient and powerful figure may owe its emergence to deeper origins than even Nietzsche was aware. Zarathustra represents a personified archetypal response to accumulated inner tensions that found their release in the kind of vatic outpouring that would traditionally be associated with religion. The high-flown language is stylistically appropriate to its archetypal source. See Jung, "Archetypes of the Collective Unconscious," 37.

15. A glance at R. D. Laing's *The Divided Self* (1960; repr., London: Penguin, 1990) or Michel Foucault's *Madness and Civilization* (1965; repr., New York: Vintage, 1988) will serve to outline the basic problems. Laing proposes (emphasis his own), "*Sanity or psychosis is tested by the degree of conjunction or disjunction between two persons where the one is sane by common consent*" (*Divided Self*, 36).

16. Lear's mental stability slips away along with his kingly authority, whereas Hamlet adheres steadfastly to a disturbing truth that the world around him is afraid to acknowledge. To the extent that Hamlet sees the madness of a world in which duplicity is expedient, he is "mad." But as a voice of moral integrity, he presents as a "sane" person in a "mad" world. In his "madness" he is witness and gadfly to the complacency of a relative sanity that masks injustice and corruption. Hamlet refuses to not see; Lear loses his "sight" (notwithstanding the "reason" in the latter's "madness"). As Nietzsche argues in *The Birth of Tragedy*, Hamlet's "nausea" followed an *increase* in knowledge and insight, not a decline in mental coherence (in *Basic Writings of Nietzsche*, 60, section 7). However, it is worth remembering a striking similar-

ity: the psychological states of both Hamlet and Lear exhibit a profound correspondence and relation to the worlds in which they occur. The moral health and order of their respective governments, the "state of the world," frames the psychological security and responses of the protagonists.

17. Laing, *Divided Self*, 39–43.

18. This application of psychological theory to a cosmological situation finds a parallel in Tarnas's *The Passion of the Western Mind*, where the modern existential predicament is compared to Bateson's "double bind" model of the schizophrenic condition ("The Post-Copernican Double Bind," 419–422).

19. For the development of the concept of the Dionysian throughout Nietzsche's overall corpus, see Adrian Del Caro, *Dionysian Aesthetics* (Frankfurt: Peter D. Lang, 1981); and Kaufmann, *Nietzsche: Philosopher, Psychologist, Antichrist*, 128–131, 168–69, and especially 281–282. Kaufmann argues that in Nietzsche's thought the concept of the Dionysian eventually came to subsume the Apollonian against which it had originally been opposed (*Nietzsche: Philosopher, Psychologist, Antichrist*, 281–282).

20. Nietzsche, *Thus Spoke Zarathustra*, 43. It bears stating here that Nietzsche himself descended into a tragic and devastating "madness" towards the end of his life. Kaufmann emphasizes medical, and specifically physical, factors in describing this episode, which lasted over ten years (*Nietzsche: Philosopher, Psychologist, Antichrist*, 67–71). While it would be a mistake to overlook the medical nature of Nietzsche's mental condition, one may see in his "madness" another sign that the ideas obsessing him were more than purely academic in their psychological implications. See note 12 above and Nietzsche's own letter to Dr. Carl Fuchs (reproduced by Kaufmann), dated July 1888, six months before his literal collapse on a street in Turin: "For just a moment put yourself into the place of one who has my *Zarathustra* on his soul" (*Nietzsche: Philosopher, Psychologist, Antichrist*, 469).

21. In many ways, Nietzsche's poetic rendering of the impact of nihilism bears comparison with Shakespeare's *Macbeth*. The radical correspondence between a loss of moral compass and disturbances in the natural order, the murder of the sovereign principle of justice, the subsequent blood-guilt, the madness, the tragic vision of the cruelty and absurdity of existence, life as being a "poor player" and "signifying nothing"—all suggest an aspect of reality that the creative sensibilities of both Shakespeare and Nietzsche were trenchant enough to incorporate.

22. Nietzsche, *Gay Science*, 279.

23. Nietzsche, *Thus Spoke Zarathustra*, 39, 52–53, 103–104, 209, and 216. The German word translated as "going down"—*untergehen*—literally "going under," and used to describe the setting of the sun, also connotes destruction and decline. See Hollingdale in *Thus Spoke Zarathustra*, 339n, and Kaufmann in *Gay Science*, 275n.

24. Nietzsche, *Thus Spoke Zarathustra*, 104.

25. See Karl Jaspers, *Nietzsche and Christianity*, trans. E. B. Ashton (Chicago: Henry Regnery, 1961), 14.

26. Nietzsche, *Will to Power*, 9, note 2. On the authority and reliability of this collection of Nietzsche's notebooks, see note 10 above.

27. Randall Havas maintains that exposing denial, both of the presence of nihilism and of the human role in the process, is central to Nietzsche's philosophical project: "It is his [Nietzsche's] most fundamental conviction that, while a commitment to truthfulness is in some way definitive of modern life, we nevertheless seek to avoid taking

responsibility for that commitment. That is to say, we fail to understand the significance of the death of God." Randall Havas, *Nietzsche's Genealogy: Nihilism and the Will to Knowledge* (New York: Cornell University Press, 1995), 1.

28. I have found a helpful theoretical framing for this cluster of paradoxes in the work of Jorge Ferrer, *Revisioning Transpersonal Theory: A Participatory Vision of Human Spirituality* (Albany: State University of New York Press, 2002). Ferrer describes the limitations of an epistemology grounded in Cartesian and Kantian assumptions for the understanding of transpersonal phenomena that by their very nature render the categories of subject-object and inner-outer radically provisional. He suggests treating transpersonal phenomena as "multilocal participatory events" instead of "intrasubjective experiences." The death of God may be understood as a "multilocal participatory event" insofar as it meets the criteria that Ferrer sets forth (116–121). Thus: a) *it is transpersonal*—it happens at a level that pertains beyond ordinary consciousness, and that has a relevance beyond any individual person; b) *it is multilocal*—it happens in the locus of a collective identity, though it may also occur for a particular individual; c) *it is participatory*—it coactively involves human participation; it may also entail an existential form of knowing that involves the total human being—body, mind, heart, and spirit; d) *it is an event*—it "happens" at a level beyond direct human control that is inflected through, but not limited to, personal experience.

29. For instance, see Friedrich Nietzsche, *Daybreak: Thoughts on the Prejudices of Morality*, trans. R. J. Hollingdale (Cambridge: Cambridge University Press, 1982), 228–29; *Gay Science*, 180–81 and 280; *Thus Spoke Zarathustra*, 109, 174–75, 181, 230–31, and 246. It is significant that Heidegger, a thinker highly sensitive to the philosophical implications of the spatiotemporal dimensions of Being, employed with precision the metaphor of a horizon. See Martin Heidegger, *Being and Time*, trans. John Macquarrie and Edward Robinson (1962; repr., Oxford: Blackwell, 2000), 65.

30. "For believe me: the secret for harvesting from existence the greatest fruitfulness and the greatest enjoyment is—to *live dangerously*! Build your cities on the slopes of Vesuvius! Send your ships into uncharted seas!" Nietzsche, *Gay Science*, 228, section 283.

31. Compare "the great noontide" in Nietzsche, *Thus Spoke Zarathustra*, 103–104, 209, 215, and 231.

32. See especially sections 108–124 and 343–347 in Nietzsche, *Gay Science*, 167–180 and 279–290. Notes 12 and 1067 from *The Will to Power* are also pertinent here: the title of note 12 is "Decline of Cosmological Values," but the succeeding paragraphs begin with the words "Nihilism as a psychological state." Nietzsche neatly summarizes three major cosmological aspects of nihilism in their relation to the human psyche: the lack of *aim*, the lack of *unity* or *totality*, and the lack of *truth*. Here also Nietzsche makes an allusion to nihilism as a post-Copernican reality: "man no longer the collaborator, let alone the center, of becoming." Nietzsche, *Will to Power*, 12–13 and 549–550.

33. Much of Nietzsche's later work may be read as an attempt to overcome nihilism through the forging of new "horizons" that did not depend upon an objective transcendent authority. In a sense, the will to power, as the motive force to self-overcoming, is a kind of transcendence from *within* the immanence of life itself (see Hollingdale, *Nietzsche*, 8). The *Übermensch* is the personification of the world-to-come rather than an afterworld. In *Thus Spoke Zarathustra*, this "transcendence within immanence" is frequently expressed in spatiotemporal language. The movement of descent, Zarathustra's down-going, the celebration of the earth and the body, and the rupture of the ground (as opposed to the biblical heavens) that "brings

to light secret powers"—all invert the traditional metaphysical framework that dis-
tributes its values from above. See Nietzsche, *Thus Spoke Zarathustra*, 214–232.

Similarly, the eternal recurrence bestows the world of becoming with the ontological status formerly reserved for other, postulated worlds. A horizon emerges from
within the phenomenal sphere. This is achieved by a reconstitution of the spatiotemporal understanding through temporality: the eternal recurrence is *the moment*
to the power of infinity.

34. John Berger, "Welcome to the Abyss," *The Guardian*, November 20, 1999 (http://
www.guardian.co.uk/books/1999/nov/20/books.guardianreview).

35. This is made explicit in section 343 of *The Gay Science*, but other sections of that
book make the same indications (for instance, sections 108, 122, 123, 129, 130,
131, and 132), not to mention the later publication of Nietzsche's *The Antichrist* (in
The Portable Nietzsche, trans. and ed. Walter Kaufmann [1954; repr., New York:
Penguin, 1976], 565–656). In *The Will to Power*, Nietzsche describes nihilism as a
necessary consequence to a moribund Christian morality (3–4 and 9–10).
Kaufmann expounds Nietzsche's anti-Christian stance in *Nietzsche: Philosopher, Psychologist, Antichrist*, 337–390.

36. Narrative versions of the Passion and Resurrection are included in all the gospels:
Matthew chapters 26–28, Mark chapters 14–16, Luke chapters 22–24, and John
chapters 18–20. Hollingdale, in his introduction to *Thus Spoke Zarathustra*, draws
out some underlying conceptual similarities between Nietzsche's ideas and Lutheran
Christianity (28–29).

37. See section 109 of *The Gay Science*, 167–169, and note 12 in *The Will to Power*, 12–
13. Richard Tarnas proposes that a "trinity of modern alienation" is represented
symbolically by the figures of Copernicus (cosmological), Descartes (ontological),
and Kant (epistemological). See Tarnas, *Passion of the Western Mind*, 418–419.

38. See Heidegger, "The Word of Nietzsche: 'God is dead'," 54 and 61.

39. See Nietzsche, *Gay Science*, 167–169 and 180–181; also *Will to Power*, 12–13 and
549–50: "*This world is the will to power—and nothing besides!*" (Nietzsche's emphasis).

40. Compare *The Gay Science*, 191–192, sec. 143, where Nietzsche argues for polythe-
ism, citing its advantages to human "horizons and perspectives."

41. In the words of Nietzsche: "The total character of the world ... is in all eternity
chaos—in the sense not of a lack of necessity but of a lack of order, arrangement,
form, beauty, wisdom, and whatever other names there are for our aesthetic anthro-
pomorphisms" (*Gay Science*, 168, section 109).

42. See Tarnas, *Passion of the Western Mind*, 304.

43. See Nietzsche, *Gay Science,* 280–283, section 344; also *Will to Power*, 7 and 10.

44. See Nietzsche, *Gay Science*, 167–173, sections 109–112, and 280–283, section 344.

45. See Nietzsche, *Gay Science, 219*, section 265: "*Ultimate skepsis.*   What are man's
truths ultimately? Merely his *irrefutable* errors."

46. See Nietzsche, *Thus Spoke Zarathustra*, 85: "'Man,' that is: the evaluator."

47. See Nietzsche, *Gay Science*, 283–285, section 345, entitled "*Morality as a problem.*"

48. For the "nakedness" of the human being when stripped of moral absolutes, see
Nietzsche, *Gay Science*, 295–296, section 352.

49. Nietzsche, *Thus Spoke Zarathustra*, 49.

50. The phrase "revaluation of all values" appears in Nietzsche's later works (especially frequently in *Ecce Homo*) and was the working title for a projected multivolume work, the first volume of which (*The Antichrist*) was the only one Nietzsche finished: see Kaufmann, *Nietzsche: Philosopher, Psychologist, Antichrist*, 110–115. For some of the positive constructions Nietzsche employed in the task of overcoming nihilism, see notes 14 and 33 above.

51. Nietzsche, *Gay Science,* 279.

Bibliography

Berger, John. "Welcome to the Abyss." *The Guardian*, November 20 1999. http://www.guardian.co.uk/books/1999/nov/20/books.guardianreview.

Del Caro, Adrian. *Dionysian Aesthetics: The Role of Destruction in Creation as Reflected in the Life and Works of Friedrich Nietzsche*. Frankfurt: Peter D. Lang, 1981.

Ferrer, Jorge. *Revisioning Transpersonal Theory: A Participatory Vision of Human Spirituality*. Albany: State University of New York Press, 2002.

Foucault, Michel. *Madness and Civilization: A History of Insanity in the Age of Reason*. 1965. Reprint, New York: Vintage, 1988.

Havas, Randall. *Nietzsche's Genealogy: Nihilism and the Will to Knowledge*. New York: Cornell University Press, 1995.

Heidegger, Martin. *Being and Time*. Translated by John Macquarrie and Edward Robinson. 1962. Reprint, Oxford: Blackwell, 2000.

———. "The Word of Nietzsche: 'God is dead'." In *The Question Concerning Technology and Other Essays*. Translated by William Lovitt. New York: Harper & Row, 1977.

Hillman, James. *Re-Visioning Psychology*. 1975. Reprint, New York: HarperPerennial 1992.

Hollingdale, R. J. *Nietzsche*. London: Routledge & Kegan Paul, 1973.

———. *Nietzsche: The Man and His Philosophy*. Louisiana: Louisiana State University, 1965.

Jaspers, Karl. *Nietzsche and Christianity.* Translated by E. B. Ashton. Chicago: Henry Regnery, 1961.

Jung, Carl Gustav. "Archetypes of the Collective Unconscious." In *The Archetypes and the Collective Unconscious.* Vol. 9, pt. 1, *The Collected Works of Carl Gustav Jung.* Translated by R. F. C. Hull. Edited by H. Read, M. Fordham, G. Adler, and W. McGuire. Bollingen Series XX. Princeton, NJ: Princeton University Press, 1969.

Kaufmann, Walter. *Nietzsche: Philosopher, Psychologist, Antichrist.* 4th ed. Princeton, NJ: Princeton University Press, 1974.

Kelly, Sean. "Space, Time and Spirit: The Analogical Imagination and the Evolution of Transpersonal Theory." *Journal of Transpersonal Psychology* 34, no. 2 (2002): 73–99.

Kuhn, Thomas S. *The Structure of Scientific Revolutions.* 1962. Reprint, Chicago: University of Chicago Press, 1996.

Laing, R. D. *The Divided Self.* 1960. Reprint, London: Penguin, 1990.

Nietzsche, Friedrich. *Basic Writings of Nietzsche.* Translated and edited by Walter Kaufmann. New York: Modern Library, 2000.

———. *Daybreak: Thoughts on the Prejudices of Morality.* Translated by R. J. Hollingdale. Cambridge: Cambridge University Press, 1982.

———. *The Gay Science.* Translated by Walter Kaufmann. New York: Vintage, 1974.

———. *The Portable Nietzsche.* Translated and edited by Walter Kaufmann. 1954. Reprint, New York: Penguin, 1976.

———. *The Will to Power.* Translated by Walter Kaufmann and R. J. Hollingdale. Edited by Walter Kaufmann. New York: Vintage, 1967.

———. *Thus Spoke Zarathustra.* Translated by R. J. Hollingdale. 1969. Reprint, London: Penguin, 2003.

Nishitani, Keiji. *The Self-Overcoming of Nihilism.* Translated by Graham Parkes with Setsuko Aihara. New York: State University of New York Press, 1990.

Swimme, Brian. *The Hidden Heart of the Cosmos.* New York: Orbis, 1996.

———. *The Universe Story: From the Primordial Flaring Forth to the Ecozoic Era.* New York: HarperCollins, 1992.

Tarnas, Richard. *Cosmos and Psyche: Intimations of a New World View.* New York: Viking, 2006.

———. *The Passion of the Western Mind: Understanding the Ideas That Have Shaped Our World View.* 1991. Reprint, New York: Ballantine, 1993.

The Dark Spirit in Nature

C. G. Jung and the Spiritual Evolution of Our Time

Keiron Le Grice

When Jung described "modern man" standing "upon a peak, or at the very edge of the world, the abyss of the future before him, above him the heavens, and below him the whole of mankind with a history that disappears in primeval mists," he was surely portraying something of himself and his own existential situation.[1] He was also characterizing the epochal nature of the time in which he lived—a time which, by all accounts, seems to have marked a critical juncture in human spiritual evolution, one that gave birth, at least in potentia, to a new relationship between spirit and nature, or the beginning of a new stage in the unfolding dialectic between them. In his contribution to this transformation, Jung can justifiably be seen as a world-historical figure, to use Hegel's term. The great person, as Jung himself points out, is one who is able to give expression to the universal constants in human experience within the unique requirements of the particular historical moment—and this is precisely what Jung did. To appreciate the significance of Jung's life, therefore, one must attempt to see it in broad historical context. Recognizing the importance of his own encounter with the unconscious psyche for the entire culture, Jung's life in many respects reflects the deepest level of expression of the archetypal conditions of his time, and this naturally makes him an especially apt figure for an archetypal astrological analysis of his life and work.

Jung was born on July 26, 1875 in Kesswil, Switzerland, at the beginning of a twenty-five-year conjunction between Neptune and Pluto, a planetary alignment heralding the onset of a new five-hundred-year cycle of radical transformation in our collective experience of and conception of the spiritual dimension of reality.[2] The Neptune archetypal principle is associated with spirituality, the divine, religion, the eternal, the transcendent, the collective unconscious, myth, fantasy, and dream, as well as problematic tendencies towards escapism, avoidance, illusion, and delusion. It is a principle that dissolves, synthesizes, spiritualizes, and sensitizes. The Pluto archetype is associated with transformation, nature, instinct, biology, and evolution. It is a principle of elemental force, intensity, extremity, and power, and is related to the mythic themes of the underworld and hell, which symbolically portray the repressed instinctual dynamism within the "underworld" of the psyche—described by the Freudian Id and the Jungian shadow archetype. The combination of these two archetypal principles is therefore especially associated with the radical transformation (Pluto) of the expression of spirit (Neptune), and with the spiritualization (Neptune) of the instincts (Pluto). A Neptune-Pluto world transit brings forth deep underlying transformations of the cultural psyche and mythic-religious visions that powerfully shape human experience and our understanding of the nature of reality. In the late nineteenth century, for example, as Richard Tarnas observes in *Cosmos and Psyche*, the Neptune-Pluto conjunction marked

> the end of an age and a transformative threshold which was symbolized in the Nietzschean transvaluation of all values, the dying of the gods that had ruled the Western spirit for two millennia and more, the subterranean dissolution of conventional Christian belief and Enlightenment assumptions, the powerful upsurge of 'the unconscious'. . . and the emergence in Western culture of a range of long-suppressed and long-developing cultural phenomena and archetypal impulses.[3]

Jung was to become a seminal figure in the further unfolding of all these phenomena during the twentieth century. To understand the significance of his life within the context of what was occurring at this time, however, we first need to reflect for a moment on the dominant trends of the previous Neptune-Pluto cycle, which began with the conjunction of 1386 to 1411 (coinciding with the start of the Italian Renaissance) and which reached the midpoint of the cycle with the

Neptune-Pluto opposition occurring between 1631 and 1660 (the start of the late modern era).[4]

The Neptune-Pluto Cycle

Focusing specifically on the spiritual dimension of experience, if one were to try to discern the major development of this period, one would, I believe, have to point to the growing conceptual and ethical separation between spirit and nature, especially as conceived within the Christian tradition. The course of human cultural and psychospiritual development during this 500-year cycle, particularly in the West, had effected a radical separation of spirit from matter, of the transcendent godhead from nature, and of human consciousness from instinct. In a number of ways, the dimensions of experience associated with Neptune and Pluto were thrust apart, even as the principles associated with these planets continued to mutually influence each other.

During this period, the element of human experience associated with Pluto was largely projected onto the figure of the Devil and excluded from the prevailing image of the divine. The Christian figure of the Devil, from the late Middle Ages (around the fourteenth century onwards, closely coincident with the start of the last but one Neptune-Pluto cycle), assumed a different character to the earlier biblical Satan, who was merely a consort of Yahweh rather than having the wholly evil and demonic character we now commonly associate with the Devil.[5] As Alan Watts observes, among the world religions "the Christian Devil is unique. No other demonic figure has ever been conceived to be so purely malicious, so sinister, and so totally opposed to the universal design."[6] Projected almost exclusively onto the figure of the Devil, Plutonic qualities—such as destruction, demonic power, ruthlessness, wrathfulness, vindictiveness, compulsion—although intrinsic to the Old Testament Yahweh, were not attributed to the person of Jesus or to the Father-God of the New Testament. Indeed, all forms of behavior and experiences connected to the instincts and nature (sex, aggression, power drives, and so forth) were concealed beneath an idealistic religious vision that excluded them from the prevailing conception of the divine. Such qualities and characteristics had nothing to do with God, it was supposed, who was conceived as all loving, all merciful, and wholly benevolent (even as His judgment cast unrepentant sinners into Hell), just as Jesus was pure and gentle, meek and mild, a celibate pacifist untainted with original sin. As Jung points out, little reflection was given among Christians as to the significance of the New Testament accounts of Jesus stealing a mule, whipping money lenders, or his instruction to be "cunning as serpents" as well as "gentle as doves." Within Christian orthodoxy, sexuality, instinct, the bodily urges and passions (relating to Pluto) were seen as inherently sinful and evil, and thus excluded by the Christian moral separation of good (light, Christ, God, spirit) from evil (dark, Devil, nature, instinct). Good and evil, God and nature, Christ and the Devil, were conceived as absolute opposites. People were taught to put themselves exclusively on the side of the good in the fight against evil. Accordingly, the Christian ethical distinction between good and evil meant that natural human instincts were often denied expression. As Friedrich Nietzsche realized, this brought a subsequent loss of life power and vital energy across a culture shaped by Christian morality. As Joseph Campbell points out, our usual understanding of the meaning of the word *demon* is illustrative of this point. Rather than signifying an evil power, the original meaning of this term, coming from the Greek *daemon*, is the dynamic of life. In the Christian tradition, the mythic and religious conceptions of the nature of reality in which the demonic is synonymous with evil put us in opposition to the life dynamic, such that Plutonic qualities of passion, desire, drive, and so forth were viewed in largely negative terms, and seen as antithetical to the spiritual life.

Simultaneously, however, the Pluto principle empowered and infused the Christian mythological imagination with its vivid portrayal of the grim fate of sinners facing an eternity in a fiery hell. Also apparent during this period was the unconscious empowerment (Pluto) of tendencies towards escapism, illusion, and avoidance (Neptune), which manifested as an excessive concern with an otherworldly afterlife in Heaven at the expense of the here-and-now, as well as the impassioned commitment to an increasingly rigid dogma espoused by Christian orthodoxy. In a sense, the Plutonic dimension of experience became cloaked behind a specious religious morality and lay simmering in the depths of the unconscious psyche, becoming increasingly charged until it was to erupt during the Neptune-Pluto conjunction and its subsequent unfolding through the twentieth century.[7]

The full implications of this separation between spirit and nature were acutely realized and experienced in the life and person of Friedrich Nietzsche. Under the Saturn-Neptune-Pluto triple conjunction of the early 1880s, a feverishly possessed and inspired Nietzsche called for the destruction of false religious ideals, the abandonment of Christianity (and all religions), and a return to the morality of the warrior caste of ancient Greece and Rome. Ruthlessly exposing what he saw as the deception and illusion fostered by the Christian image of the divine (a theme expressive of the Saturn-Neptune complex, driven by Plutonic intensity), Nietzsche effectively stood in judgment on a religious world conception based on the separation of spirit from nature, which had led to a repression of the instincts and the denigration of nature.[8] Stepping beyond the established religious and moral psychological boundaries of the modern West, Nietzsche experienced a volcanic eruption of the unconscious power of nature within his own psyche.

Freud and the early pioneers of depth psychology later assumed the urgent task of raising to conscious awareness the repressed instinctual basis of human experience through the more systematic approach of psychoanalysis. The repressed Plutonic dimension of experience was thus brought back into the foreground of human experience. Depth psychology confronted people with their unconscious desires, power drives, frustrated sexual impulses, and more—the lid was taken off the seething cauldron of the instincts. Much of what was considered evil, demonic, primitive, and uncivilized in human nature was to be faced immediately and directly as a personal psychological reality.

As we will see, Jung was a key figure in this development, attempting to bring an awareness of the dark, "shadow" elements of human experience into the Christian conception of the godhead. Beyond this, by integrating the instinctual dynamism and evolutionary power of nature into modern psychology and spirituality, Jung's life and work were especially relevant to other major themes associated with the Neptune-Pluto archetypal combination. In its exploration of the shadow and the depths of human experience, and in its emphasis of radical psychological transformation, Jungian psychology is Plutonic in character; in its exploration of the numinous, myth, religion, dreams, and fantasies, it is Neptunian in character. Similar emphases can also be found in others born during the Neptune-Pluto conjunction, notably Sri Aurobindo and

Pierre Teilhard de Chardin, whose own work brought together spirituality and evolution, providing a deeper understanding of the process of psychospiritual transformation.

Planetary Alignments at Jung's Birth

Although in a wide orb at Jung's birth, the outer-planet alignment of Neptune and Pluto had a greater relevance to Jung's life than it might have had otherwise because of its relationship to the major aspects in his chart. These aspects include the Sun square Neptune, which suggests the propensity to be particularly consciously attuned to and challenged by the spiritual dimension of experience; Saturn square Pluto, which, in Jung's case, relates to the Herculean labor of the transformation of the psyche through individuation; and Saturn and Pluto in a T-square configuration with Uranus, with this larger aspect pattern suggesting the sudden crisis of transformation Jung was to experience in his life—a topic we will consider shortly. Jung also had the Moon, in a conjunction with Pluto and close to exact square to Uranus, positioned approximately midway between the approaching Neptune-Pluto conjunction, suggesting a natural emotional attunement to themes associated with this world transit alignment. And his natal Uranus square Neptune relates most especially to the sudden revelatory shifts of consciousness and insights that were to characterize Jung's life and work. Placed within the particular historical context of late-nineteenth century mainland Europe, it is this complex configuration of archetypal factors, finding expression through Jung's highly developed intellectual and spiritual sensibility, that, together with innumerable other factors—biographical, cultural, and more—has decisively shaped the contemporary understanding of and approach to spirituality and psychological transformation from the mid-twentieth century onwards, especially with the promulgation of Jung's ideas in recent decades. The expression of all these archetypal complexes in Jung's life was shaped at its deepest level by the archetypal dynamics associated with the Neptune-Pluto conjunction.

Figure 1 **Birth Chart of C. G. Jung**

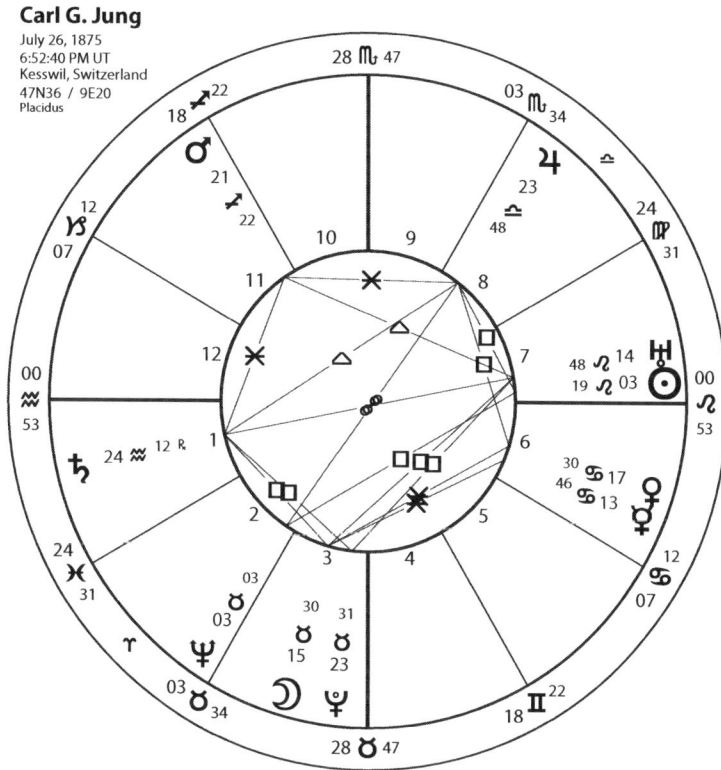

Carl G. Jung
July 26, 1875
6:52:40 PM UT
Kesswil, Switzerland
47N36 / 9E20
Placidus

In the pages to follow, I will briefly examine three decisive biographical periods in Jung's life that are powerfully illustrative of some of the major archetypal complexes in his chart and expressive of the archetypal relationship between Neptune and Pluto, including Jung's childhood, the years of Jung's confrontation with the unconscious between 1912 and 1918, and the years of his major alchemical studies in the 1940s and 1950s. Considering these periods, it is possible to observe a diachronic sequencing of qualitatively similar archetypal themes, and also an evolutionary development in the expression of the archetypal complexes during the course of Jung's life as he worked through the various types of transformative experience associated with these complexes.[9] While focusing primarily on the relevance of Jung's life for the collective spiritual transformation associated with the Neptune-Pluto conjunction, this analysis will also examine how this archetypal complex finds specific expression in Jung's experience by considering the natal alignments in his chart and, to a lesser extent, his personal transits.

Jung's Childhood Dreams and Experiences

One is struck, reading Jung's autobiography, *Memories, Dreams, Reflections,* by the extent to which his life's work was prefigured and initiated in his early childhood dreams. The first dream, the earliest Jung could remember, occurred when he was between three and four years old:

> Suddenly I discovered a dark, rectangular, stone-lined hole in the ground. I had never seen it before. I ran curiously forward and peered down into it. Then I saw a stone stairway leading down. Hesitantly and fearfully, I descended . . . I saw before me in the dim light a rectangular chamber about thirty feet long. The ceiling was arched and of hewn stone. The floor was laid with flagstones, and in the centre a red carpet ran from the entrance to a low platform. On this platform stood a wonderfully rich golden throne . . . a magnificent throne, a real king's throne in a fairy tale. Something was standing on it which I thought at first was a tree trunk twelve to fifteen feet high and about one and a half to two feet thick. It was a huge thing reaching almost to the ceiling. But it was of a curious composition: it was made of skin and naked flesh, and on top there was something like a rounded head with no face and no hair. On the very top of the head was a single eye, gazing motionlessly upwards.[10]

Jung recalled being "paralyzed with terror" when confronted by the enormous, worm-like "man-eater," as his mother described it within the dream. He realized, much later in his life, that the creature was a giant phallus, a "ritual phallus," which, in Jung's view, was a compensatory image produced by the unconscious, counterbalancing the prevailing conception of an all-loving, all-good "Lord Jesus." The phallus was a symbol of a "dark Lord Jesus," representing qualities many of his contemporaries were not willing to see in the Savior or in God; and the dream was also, Jung realized, a portrayal of the cannibalistic symbolism of the Christian Mass.

Here, then, at the start of the Neptune-Pluto conjunction around 1880, came a vision that appeared to depict some kind of reactivation of a subterranean divinity, a powerful expression of the dark spirit of nature associated with Neptune-Pluto. This was the sexual, instinctual, chthonic, devouring principle associated with Pluto assuming imaginal form. Occurring around the same time as Nietzsche's *Thus Spoke Zarathustra,* which was addressing related themes, here was the repressed and ignored spiritual-instinctual power of nature graphically impressing itself on the mythic vision of the contemporary psyche.

Through this dream, Jung believed that he had been "initiated into the secrets of the earth," and entered "the realm of darkness." "What happened then," he explains, "was a kind of burial in the earth, and many years were to pass before I came out again."[11] The motifs of this dream, and its consequences, clearly reflect the particular combination of Moon-Saturn-Pluto in Jung's chart: the underground setting, the phallic monster, the descent, and the transformative initiation point to the archetypal Pluto; the stone chamber, the burial, and the entombment relate to Saturn; and the role of his mother and the fact the dream occurred in childhood are connected with the archetype of the Moon. The Saturn-Pluto combination also relates to the fateful quality of the dream and the experience of extreme fear, with the instinctual force of Pluto empowering the self-defensive, fearful responses associated with Saturn. The Saturn archetype is related to the ego-structure, and to the limitations, restrictions, and the often acute sense of separation that comes with ego-consciousness. Under the archetypal influence of Pluto, these characteristics are typically intensified and driven to an extreme. Unsurprisingly, then, the entire mood of Jung's life at the time was oppressive and foreboding. A fear of suffocation and an "unbreathable" atmosphere in the house gripped his waking hours; it was a time of dark, unspeakable secrets, and of overwhelming compulsions.[12]

The Moon-Saturn-Pluto alignment in Jung's birth chart gives us more precise information as to how he experienced the archetypal energies accompanying the Neptune-Pluto world transit, pointing in particular to the decidedly Saturnian nature of his experience of the Plutonic underworld of the psyche. The moral imperatives and boundaries (associated with Saturn) that had rendered the Plutonic dimension of experience taboo across the wider culture in the first place were instilled in Jung through his upbringing in the pious Protestant context in which he lived. Thus, to experience the dark power of nature through his childhood dreams and visions was to struggle against the morality of his time, against his own ingrained sense of good and evil, against the expectancy of divine punishment. To access experiences associated with the archetypal Pluto, Jung was to encounter great inner resistance and deep fear (Saturn). As he wrote later in his life:

> In order to grasp the fantasies that were stirring within me "underground," I knew that I had to let myself plummet down into them, as it were. I felt not only violent resistance to this, but a distinct fear.[13]

Yet the very extremity of the struggle was essential to Jung's realization. He had to hold both poles within him: the established morality of Western culture and his direct experience of the unrecognized subterranean divinity—a power which is "beyond good and evil" and which thus requires the moral discernment of a developed ego if one is to relate to it in a balanced and constructive manner.

A second experience from Jung's childhood further develops these themes. Again, contravening the moral conditioning of his childhood, Jung found himself compelled to entertain what seemed to him to be a blasphemous, sinful thought:

> I gathered all my courage, as though I were about to leap forthwith into hell-fire, and let the thought come. I saw before me the cathedral, the blue sky. God sits on His golden throne, high above the world—and from under the throne an enormous turd falls upon the sparkling new roof, shatters it, and breaks the walls of the cathedral asunder.[14]

The divine feces suggests something rejected, forced down into the bowels of the Earth or the underworld, some base elemental force crashing into the daylight "conscious" view of God's cathedral. Those aspects of the divine that had been overlooked were first to destroy and then to begin to regenerate the idealistic Christian spiritual vision and its metaphysical foundations. Just as for Nietzsche the Dionysian principle of impassioned frenzy and self-annihilation was the essential compensatory counterforce to the ordered beauty and control of the Apollonian sensibility and to Christian morality, here in Jung's vision was the dark, destructive power of God in nature, God in base, instinctual, unregenerate form, shattering the pristine Christian vision of the glory of the Almighty, seated on His throne in the sky—a theme that was still preoccupying Jung in his *Answer to Job* in the 1950s.[15] For the Pluto-Hades-Dionysus-Devil principle, largely excluded from the cultural imagination of the divine and therefore denied legitimate expression within personal experience during the previous centuries, was to unconsciously possess the culture and the psyche that denied its reality and that preferred to rest in the inadequate, childlike vision of a wholly benevolent deity, devoid of sin or moral ambiguity. The Plutonic dimension of experience—the biological, primitive, elemental, unconscious, instinctual, and sexual—was powerfully re-impressing itself on the Neptunian spiritual vision of late-nineteenth century and early twentieth century Western culture.[16] In the lineage of Goethe and Nietzsche, Jung was to be a pivotal figure in this reawakening.

The Confrontation with the Unconscious: 1912–1918

Jung's early visions anticipated the major themes and problems he was to work through in the future, but it was during his mid-to-late thirties that he was to dramatically revisit the underworld scenes of his early years. The "confrontation with the unconscious," as he termed it, was triggered by the much-discussed break from Freud and the Psychoanalytic Association following the publication of his *Wandlungen und Symbole der Libido* in 1912 in which Jung presented his own interpretation of incest theory that emphasized the mythological, rather than the sexual, aetiology of incest fantasy. Cast adrift from his former intellectual and professional moorings following this break, Jung was thrown back onto himself and entered a state of deep introspection and existential turmoil—a psychospiritual crisis, as it might be called today.

Particularly relevant to understanding the archetypal dynamics of this crisis is the Saturn-Uranus-Pluto T-square in Jung's natal chart. Uranus is the archetypal principle associated with going one's own way in life, with expressing one's own individual genius. The Saturn principle is associated with the established order, protective and prohibitive boundaries, the status quo, authority, tradition, supporting structures, worldly roles, and responsibilities. It marks the threshold of what is known, accepted, and already established. In Jung's case, we can see that it was the impetus to advance his own theories rather than uphold the established psychoanalytic perspective that thrust him into his crisis period. Contravening the patriarchal authority of Freud, Jung chose to remain true to his own insights and experiences, and this meant professional and, to a certain extent, psychological isolation: "When I parted from Freud, I knew that I was plunging into the unknown. Beyond Freud . . . I knew nothing; but I had taken a step into darkness."[17]

In archetypal terms, Saturn-Uranus is particularly associated with the sudden crisis, when the problematic aspects of one's existence—one's fears, weaknesses, and any sense of inferiority—can be sharply and often suddenly accentuated, brought to a head in acute crises. This combination is associated with the experience of the "break"—whether breaking free, breaking through, or breaking down. Because Saturn and Uranus are in major aspect to Pluto in Jung's chart, this separation from the psychoanalytic movement was inextricably connected with the underworld descent and radical psychological transformation associated with Pluto, as well as instigating the ongoing power struggle with Freud.

By December 12th of 1913, Jung's crisis had reached a decisive moment:

> I was sitting at my desk once more, thinking over my fears. Then I let myself drop. Suddenly it was as though the ground literally gave way beneath my feet, and I plunged down into dark depths.[18]

Both the sudden breakthrough against resistance and the unexpected fall into the depths are clearly related to the Saturn-Uranus-Pluto T-square. In Jung's case, this moment seems to have marked the sudden collapse of the ego-structure and its defenses, accompanied by the breaking through of unconscious contents into consciousness through an exceptionally vivid fantasy sequence. The Uranus archetypal principle, as Stephen Arroyo points out, relates to that which lies just below the surface of consciousness and is ready to emerge into awareness.[19] It is the principle that activates, emancipates, and releases; and in this case what was liberated in Jung was the Plutonic experience of the depths underlying normal ego-consciousness. The dark spirit of nature, the "spirit of the depths," as he called it in his own journals, irrupted into Jung's consciousness.

Following this descent, Jung faced an immense struggle to discipline and control the activated instincts, primitive drives, and undifferentiated emotions associated with Pluto. "I was frequently so wrought up," he explained, "that I had to do certain yoga exercises to hold my emotions in check."[20] However, Jung's concern was not principally to control and suppress his emotions, desires, and instincts, but rather to bring to consciousness the underlying motivations behind them in order that they might be transformed and integrated. "As soon as I had the feeling I was myself again, I abandoned this restraint on the emotions."[21] This oscillation between restraint, control, and discipline on the one hand, and the abandonment of restraint, giving free reign to the instincts and emotions, on the other hand, again reflects the archetypal configuration in Jung's chart involving the Moon (emotions), Saturn (restraint, control), Uranus (freedom, liberation, breakthrough), and Pluto (the instincts, penetrating, depth, underlying motivations).[22]

Jung's journey through the underworld of the psyche pulled his attention away from the daylight realm of ordinary concerns into a world of visions, personified archetypal figures, dark oppressive moods, and fateful instructive dreams. Analyzing Jung's experiences during these years, one can again recognize the interconnection between the Neptune-Pluto world transit cycle under which he was born and the archetypal dynamics of his birth chart. During this time, Jung's consciousness was subject to an influx of streams of fantasy images infused with the dynamism and power of the instincts that had been repressed during the long course of ego-development, especially during the previous Neptune-Pluto cycle. Jung would later describe this experience, in the language of his own psychological theories, as a compensatory response by the unconscious to the one-sidedness of ego-consciousness, a response that was experienced as an "avenging deluge" in which the flow of instinctual power and imagery from the collective unconscious threatened to utterly overwhelm him.[23] With this deluge came the danger of self-loss, of dissolution in the collective psyche, or unconscious possession and psychosis—experiences that are directly expressive of the Neptune-Pluto complex, bringing together themes of flow, water, imagery, fantasy, dissolution, and unreality (relating to Neptune) with the power, instincts, unconscious compulsion, and states of possession (relating to Pluto). Together Neptune and Pluto can be experienced as ceaseless turmoil, as a maelstrom of emotions, images, visions, and drives.

To give more context, this was the same Neptune-Pluto conjunction present in the birth charts of a number of immensely influential figures whose lives and deeds have utterly transformed the modern world and world view. The Neptune-Pluto complex empowered the mythic imagination of J. R. R. Tolkien to bring forth *The Lord of the Rings* with its gripping vision of an enchanted world and its telling of the mythic adventures of Frodo Baggins, Gandalf, and their companions on the epic journey to destroy the Ring of Power. It was the same archetypal complex that forged the imaginative capacities of Walt Disney to create the fantastic adventures of Mickey Mouse and other inhabitants of the "Magic Kingdom," transmitted via his cartoon empire to generation after generation of children and adults alike. In physics, a similar empowerment of the imagination helped Albert Einstein re-envisage the nature of material reality with his Special and General Theories of Relativity, transforming his field much as Pablo Picasso, himself born during the Neptune-Pluto conjunction, had done in painting. As we have seen, the Neptune-Pluto combination also found expression in the evolutionary visions of Sri Aurobindo and Pierre Teilhard de Chardin (both also born during the conjunction) in which spirituality was recognized to be present within the natural world impelling human beings to evolve to ever greater levels of consciousness and self-realization. In shadow form, however, the same energies, present in the birth charts of Adolf Hitler and his contemporaries, were the driving archetypal factors behind the pathological ideology and mythos of Nazism and Arian supremacy (drawing in part on a warped interpretation of Nietzsche's ideal of the *Übermensch*), which became fully manifest in the collective psychosis of the Second World War. This is the subject of Jung's essay on the Germanic god Wotan, a mythic personification of some of the characteristics associated with the archetypal Pluto. For better or for worse, then, in each of these examples, the human imagination became intensified and empowered to bring forth stirring new visions, fantasies, dreams, and creative works, which in some cases were populated with Plutonic themes such as extreme ordeals, fiery hells and underworlds, possession states, monsters of the depths, rebirth and renewal, destruction and creation, and transformation and evolution.

The following passage is an excellent account of Jung's own experience of the Neptune-Pluto complex, mediated by his natal configuration of Moon-Saturn-Uranus-Pluto:

> An incessant stream of fantasies had been released, and I did my best not to lose my head but to find some way to understand these strange things. . . . I was living in a constant state of tension; often I felt as if gigantic blocks of stone were tumbling down upon me. One thunderstorm followed another. My enduring these storms was a question of brute strength. Others have been shattered by them—Nietzsche, and Hölderlin, and many others. But there was a demonic strength in me. . . . When I endured these assaults of the unconscious I had an unswerving conviction that I was obeying a higher will, and that feeling continued to uphold me until I had mastered the task.[24]

Jung's description of living under a "constant inner pressure" and being possessed of "a demonic strength" are themes associated with Saturn-Pluto. The Saturn-Pluto complex is typically associated with the struggle for self-control, the fortification of the will, the exercising of ego-strength, and the confrontation with or harnessing of demonic forces. Cultivating these qualities was imperative if Jung were to be able to contain the flow of drives and fantasies from the unconscious without being torn to pieces—a fate that, as Jung knew only too well, had befallen Nietzsche some thirty years earlier. Jung, however, was blessed with more favorable life circumstances than Nietzsche in that Jung had a wife and family and emotional commitments to his patients (Moon) that provided the necessary connection to consensus reality (Saturn) to anchor him during his descent into the underworld (Pluto). Simultaneously, Jung struggled to assert his own conscious identity and will (the Sun) against the disorienting deluge of images that threatened to overwhelm him (Neptune), repeating to himself his name, address, profession, and other details that would reinforce his personal identity against the threat of dissolution in the transpersonal psyche.[25]

With regard to personal transits, from an archetypal perspective the entire episode seems to have been triggered by transiting Uranus forming an opposition to Jung's natal Sun and square to his natal Neptune. True to its established meaning as a principle of awakening, liberation, creative change, and disruption, the Uranus principle seems to have been the catalyst for the influx of the stream of fantasy images (Neptune), and for the resultant pervasive sense of disorientation and identity confusion associated with the natal configuration of Sun-Neptune in Jung's chart. One could say that during the Uranus transit to the Sun-Neptune alignment the dam burst, as it were, leading to the sudden influx of the unconscious, calling forth Jung's creative genius for pioneering a new approach to interpreting the visions and fantasies produced by the unconscious. In his descriptions of this period of his life as a "state of disorientation," "inner uncertainty," and of being "suspended in mid-air," Jung captured the essence of a dominant theme of the Uranus-Neptune complex, which is associated with just this kind of pervasive uncertainty and disorientation arising from the jolting, disruptive effects of sudden glimpses into subtler dimensions of reality and the experience of non-ordinary states of consciousness.[26]

In the labyrinth of fantasy images and intoxicating states of consciousness Jung encountered, one can easily lose one's mind or one's grip on reality. To maintain a sense of control and a measure of detachment is therefore supremely challenging. "Only by extreme effort," Jung confessed, "was I finally able to escape from the labyrinth."[27] But emerge he did, and on later reflection Jung came to realize that his period of inner turmoil and struggle, for all its difficulties and dangers, had provided the "*prima materia* for a life's work."[28]

This period of Jung's life, through until 1930, has recently been illuminated in great profundity with the publication of *The Red Book*, containing records of Jung's fantasy sequences, active imaginations, and dialogues with the unconscious during these years. The book is a quite astonishing revelation of the dark mystery of the divine, among the finest recorded testimonies to the spiritual transformation associated with the Neptune-Pluto conjunction. Archetypally, the "spirit of the depths," with whom Jung is engaged in ongoing conversation, perfectly expresses both Neptune (spirit) and Pluto (depth). Confounding the Christian image of a wholly benevolent deity, it is a revelation of the morally ambiguous nature of God, and of the gods (now conceived as archetypes), demonstrating the inextricable connection between good and evil, light and shadow, beatific divine love and terrible divine power. Fundamentally, the dimension of God that Jung is most engaged with throughout *The Red Book* is unmistakably Plutonic in nature. In the Appendix, for instance, Jung portrays the mysterious "God of the cosmos," the frightful Abraxas, who displays a litany of characteristics associated with Pluto, exemplified by the following passage:[29]

> He is the God of the cosmos, extremely powerful and fearful. He is the creative drive, he is form and formation, just as much matter and force. . . . He tears away souls and casts them into procreation. He is the creative and the created. He is the god who always renews himself, in days, in months, in years, in human life, in ages, in peoples, in the living, in heavenly bodies. He compels, he is unsparing. If you worship him, you increase his power over you. Thereby it becomes unbearable.[30]

We can recognize here, furthermore, that the particular inflection given to the character of the Plutonic god Abraxas is again Saturnian, no doubt reflecting the larger Saturn-Moon-Pluto complex in Jung's natal chart. Pluto relates specifically to the Shiva-like creative, destructive, and procreative power of Abraxas, to the god's capacity for inexorable renewal, and to his compelling ruthless power and extremity of expression. Saturn is connected to form, the measurements of time, and to fear, and also contributes to Abraxas' remorseless, unsparing character and inescapable influence, which are directed towards the soul (the Moon).

In another passage, Jung is led by the spirit of the depths to a radically new understanding of the nature of soul:

> The spirit of the depths considers the soul . . . as a living and self-existing being, and with this he contradicts the spirit of this time for whom the soul is a thing dependent on man, which lets herself be judged and arranged, and whose circumference we can grasp. I had to accept that what I had previously called my soul was not at all my soul but a dead system.[31]

"The spirit of the depths," he adds, "forced me to speak to my soul, to call upon her as a living and self-existing being."[32] In terms of the universal archetypal factors shaping Jung's experience, we can see in these passages the Saturnian "dead system" and "judgment" of the soul that was unexpectedly contradicted by the encounter with the spirit of the depths.

Awakening to the soullessness of modern life, it was impressed upon Jung that abstinence from the world, a turning within to face the depths of the psyche, and a firm control of the passions were unavoidable requirements if he were to rediscover his soul:

> He whose desire turns away from the outer things, reaches the place of the soul. If he does not find the soul, the horror of emptiness will overcome him, and fear will drive him with a whip lashing time and again in a desperate endeavor and a blind desire for the hollow things of the world. He becomes a fool through his endless desire, and forgets the way of his soul, never to find her again. . . . My friend, it is wise to nourish the soul, otherwise you will breed dragons and devils in your heart.[33]

Jung describes here some of the major challenges and dangers that are intrinsic to the experience of the Pluto archetype in association with Saturn: throwing oneself blindly into worldly commitments and hard work to which one is not suited, being ceaselessly driven by one's passion to the point of suffering, and being scorched by the fire of desire to the loss of one's humanity. With the Saturn-Pluto combination, the Plutonic instinctual drive and passion can feel ines capable, relentless, crushing, or painfully inhibited.

One could give example after example of the Plutonic nature of the spiritual revelation contained in *The Red Book*. Virtually every page reflects the mystery and power of the dark, chthonic side of the divine, pressing to make itself known to the consciousness of modern humans. Unsurprisingly, as we will now consider, the same themes were equally evident in Jung's central spiritual and professional concerns in the latter part of his life.

Alchemical Studies During the 1940s and 1950s

By the mid-1940s, when Jung's attention turned increasingly towards the study and interpretation of the arcane texts of alchemy, Neptune and Pluto had moved into the beginning of a long sextile (60-degree) alignment. This particular angular relationship, in keeping with its established meaning in astrology, brought with it an increasing sense of perspective on the geyser-like eruptions of the collective unconscious occurring during the previous conjunction.[34] If the conjunction had marked the bursting forth of new spiritual visions and powerful experiential manifestations of the dark spirit of nature, the sextile permitted an emerging insight, after the Second World War, into what had taken place during the conjunction—an open invitation, as it were, to explore and better understand in a more stable climate the relationship between these archetypal principles.

In this new archetypal context in the mid-twentieth century, Jung was able to work out the relationship in his own life between the Sun-Neptune and the Moon-Saturn-Pluto complexes in his chart. The symbolism of alchemy provided a language and a method to enable him to do this.[35] The relationship between the alchemical *Sol* and *Luna* was especially relevant as a symbol of the confrontation between ego-consciousness and its unconscious ground. As Jung explains:

> This confrontation is expressed, in the alchemical myth of the king, as the collision of the masculine, spiritual father-world ruled over by King *Sol* with the feminine, chthonic mother-world symbolized by the *aqua permanens* or by the chaos.[36]

The spiritual-father world of the alchemical king is a fitting symbol of Jung's natal Sun-Neptune combination, with the Sun as the father-king of the Neptunian spiritual dimension (one might recall here Jung's dream of the fairytale king's throne). The "chthonic mother-world," on the other hand, is a symbolic expression of the Moon-Pluto combination in Jung's chart, reflecting the Moon's association with the mother and the matrix of being and Pluto's association with the primitive, instinctual dynamism of the unconscious ground of existence. As Jung demonstrates, it is an essential element of the work of individuation to bring *Sol* and *Luna* into relationship, a process that is portrayed through the various transformative operations performed by the alchemists. "It is the moral task of alchemy," Jung proclaims, "to bring the feminine, maternal background of the masculine psyche, seething with passions, into harmony with the principle of the spirit—truly a labor of Hercules!"[37] The moral task is Saturnian in essence; the seething passions and maternal background to the psyche relate to Moon-Pluto; Neptune is the "principle of the spirit"; and the Herculean labor of transformation relates to Saturn-Pluto—the arduous work to be done in the underworld, containing and controlling the instincts and passions that they might then be transformed.

The goal of the alchemical process is to unite the transcendent spiritual principle with "the dark chthonic aspect of nature," which, Jung explains, "is not only the darkness of the animal sphere, but rather a spiritual nature or a natural spirit."[38] This insight relates directly to the Neptune-Pluto alignment as it was finding expression through Jung. In alchemy, the chthonic spirit is also known as the *anima mundi*—the world soul—that is thought to lie trapped within matter, caught up in the processes of nature, "imprisoned in the chains of *physis*."[39] The transcendent principle is the "counsel of the spirit" that calls the soul to awaken from her slumber in matter and imposes upon the conscious ego the task of actualizing the soul's release.[40]

Another mythic figure that reflects the new image of the divine, emerging at the turn of the twentieth century with the Neptune-Pluto conjunction, was the alchemical god-man *Mercurius* who, as a personification of the unconscious, is the central figure in the entire alchemical drama, both the subject and object of the alchemical work of transformation. In his extensive discussions of *Mercurius* scattered through his three volumes on alchemy, Jung provides a number of synonyms and associations that point to the fundamentally Plutonic character of the god, inflected as ever by the archetypal qualities pertaining to Saturn and the Moon: *Mercurius* is the *spiritus vegetavius*, the chthonic living spirit in matter and yet, paradoxically, the spirit which keeps the soul imprisoned therein[41]; he is the *unus mundus*—the original state of unconscious wholeness, the "undifferentiated unity of the world"; he is the king's instinctual animal side that is encountered in the descent into the realm of the unconscious and is the power that "darkens the sun"[42] and produces the *sol niger*—the black sun in the Earth[43]; *Mercurius* is also related to the anima, which, like the alchemical *Luna*, is to be transformed through its differentiation in the light of consciousness; he is also described as a "God Man," "the Son of God," and the "Anthropos,"[44]; finally, *Mercurius* is morally ambivalent, bright and dark, male and female, and is "beyond good and evil" as he represents the collective unconscious where such moral categories do not apply.[45] He is, as the aforementioned examples make clear, a symbol that embraces and reconciles all pairs of opposites.

In the alchemical transformation process, *Mercurius* is also represented by the serpent—a classic theriomorphic symbol of the Pluto principle. The alchemical work seeks to effect a transformation of the serpent into the *lapis philosophorum* (the philosopher's stone, which is the goal of the alchemical process) via a number of operations performed in the *vas alembic*. The alembic functions as a kind of uterus of spiritual rebirth: it is a vessel of transformation in which, through heating, poisonous impurities are eliminated. Jung elaborates:

> Through the incubation [in the vessel] the snake-like content is vapourized, literally 'sublimated,' which amounts to saying that it is recognized and made an object of conscious discrimination.[46]

Poison is often represented by sulphur, which, according to Jung, is understood by the alchemists to constitute "the inner fire,"[47] which is to be found "in the depths of the nature of *Mercurius*."[48] Sulphur, we learn, is "evil smelling" and as a poisonous substance has the power to corrupt and to "blacken the sun."[49] As the fire of *Mercurius*, Jung therefore sees sulphur as corresponding to the "unconscious dynamism and compulsion" (Pluto) that thwarts the conscious will and thus forces the conscious ego to turn its attention to the unconscious.[50] In this way, sulphur is responsible for bringing about an expansion of consciousness, justifying its supposed identity with Lucifer, "the bringer of light."[51]

The serpent is one among many theriomorphic symbols of the transformation of *Mercurius* that occur during the alchemical opus. As Jung explains:

> the union of consciousness (Sol) with its feminine counterpart the unconscious (Luna) has undesirable results to begin with: it produces poisonous animals such as the dragon, serpent, scorpion, basilisk, and toad; then the lion, bear, wolf, dog, and finally the eagle and the raven. The first to appear are the cold-blooded animals, then the warm-blooded predators, and lastly birds of prey or ill-omened scavengers.[52]

The animal symbols represent the "dangerous preliminary stages" of the encounter with the unconscious, and the sequence of transformations into different animal forms reveals, in Jungian terminology, increasing degrees of conscious differentiation of the anima.[53] With each transformation, the primitive drives and instincts become progressively more tamed, more humanized.

As will be clear, this sequence of transformations, and the alchemical process in general, fits, with remarkable precision, archetypal themes associated with the Moon-Saturn-Pluto alignment in Jung's birth chart. Pluto is represented by the serpent *Mercurius* and the animal symbolism of the unconscious; it relates to unconscious compulsion and the challenge of overcoming such compulsion through the burning out of impurities. The archetypal Pluto finds expression as the purifying fire of transformation. Saturn is associated with the containing structure of the alchemical vessel, the labor and discipline of the process, and the base material that is transformed into spiritual gold. The Moon relates to the anima, the emotions, and the soul.

Ultimately, the aim of the alchemical opus is to liberate (Uranus) the soul (the Moon) from the chains of *physis* (Saturn). This goal may be achieved by the overcoming of desirousness, instinct, and compulsive grasping (relating to Pluto). When the soul is no longer subject to domination by the appetites, it may become a vehicle for the realization of the *anima mundi*, which is envisioned as the "slumbering spirit" of nature. The emancipation of the *anima mundi* from matter, and the separation of the soul from the appetites and desires of the body, is to be effected through a process known as *separatio* (or *distractio*).[54] By a withdrawal of projections, ascetic self-denial, and "the careful investigation of desires and their motives," it is possible to bring to an end the ego's "unconscious identity with the object" and grow free of "the turbulence of the emotions."[55] The natural state of *unio naturalis*, as the alchemists called it, in which the soul is inextricably bound up with the bodily sphere, is brought to an end by liberating the soul from the body, and then effecting a union of the rational-spiritual principle with the soul. This union is known as the *unio mentalis*, the completion of the arduous first stage of the alchemical *coniunctio* of *Sol* and *Luna*, resulting in deep self-knowledge and producing "a realistic and more or less non-illusory view of the outside world."[56]

Again, in archetypal terms this process well describes the dynamics of Jung's Moon-Saturn-Uranus-Pluto complex: the unconscious identity of the soul and the bodily instincts refers to the initial expression of the

Moon-Pluto complex, with Saturnian discipline and denial used to overcome emotional turbulence (Moon-Uranus-Pluto), and to bring to an end (Saturn) the initial state of unconscious identity with the instincts (Moon-Pluto). In these ways and more, this passage, and Jung's alchemical studies in general, describe the working out of the complexes in Jung's personality on a scale that was of significance not only to him personally, but to the cultural zeitgeist of the time and to the overarching evolutionary trajectory of the modern spirit. Through his alchemical investigations, Jung was actualizing one of the primary challenges of the unfolding Neptune-Pluto cycle, bringing together the Christian recognition of a transcendent divinity with the repressed and largely unrecognized dark spirit of nature.

The alchemical work and the process of individuation reach their fulfillment with the *coniunctio* of the conscious ego with its unconscious ground, personified as the union of *Sol* and *Luna* or of *Rex* and *Regina*. "Individuation," Jung explains, "is a 'mysterium coniunctionis,' the self being experienced as a nuptial union of opposite halves."[57] The Self as the new center of the psyche is realized, born into consciousness, from the "chymical marriage," of *Sol* and *Luna*.[58] Jung identifies three stages of this alchemical conjunction. The first, as mentioned, is the *unio mentalis*—the union of the rational-spiritual perspective (associated with ego-consciousness) with the soul, liberated from the unconscious world of matter and the body. The two further stages are the reunion of the separated *unio mentalis* with the body—a "reanimation" of the body[59]—and finally, the *mysterium coniunctionis* or *unio mystica*, which is tantamount, according to Jung, to a union of the individual with "the eternal Ground of all empirical being."[60] Here the alchemical work has reached its final end of triumphant rejuvenation and the union of nature (Pluto) and spirit (Neptune):

> It expresses, psychologically, the joys of life and the life urge which overcome and eliminate everything dark and inhibiting. Where spring-like joy and expectation reign, spirit can embrace nature and nature spirit.[61]

Jung's explorations of alchemy, then, brought to some kind of conclusion the themes of his childhood experiences and the labor of transformation imposed upon him since the time of his "confrontation with the unconscious." Through his interpretation of the psychological significance of the alchemical texts, Jung helped to bring to consciousness and articulate for modern culture the deeper meaning of the Neptune-Pluto conjunction of the late-nineteenth century.

Looking back now, it seems to be the case that this world transit alignment marked the beginning of a new cycle of human psychospiritual evolution. This cycle is bringing forth, among other things, a spiritualization of the instincts—not through their repression, as before, but now through the integration of the instinctual power of the unconscious with consciousness, centered on the rational ego.[62] Through the dialectic between ego and unconscious during individuation, the modern human can participate in the process of the realization of the dark spirit of nature, helping to make this known in the light of human self-reflective consciousness, thereby contributing to the progressive evolution of the divine.

Notes

1. Carl Gustav Jung, "The Spiritual Problem of Modern Man." In *Civilization in Transition*, 2nd edition, volume 10 of *The Collected Works of C. G. Jung*, translated by R. F. C. Hull (Princeton: Princeton University Press, 1989), 196–197.

2. The Neptune-Pluto cycle is discussed by Richard Tarnas in *Cosmos and Psyche: Intimations of a New World View* (New York: Viking, 2006), 409–418. Following Tarnas, I am using an orb of twenty degrees for the conjunction between these planets, which forms part of the longest of all the planetary cycles currently studied in astrology. In practice, the overall operative range of this transit seems to extend from something like 1875 through 1918, partly because the Neptune-Pluto conjunction also coincides and overlaps with the Uranus-Pluto opposition of 1896–1907 and the Uranus-Neptune conjunction of 1899–1918, lending it added significance. Periods of history in which the three outer planets are simultaneously in axial alignments are extremely rare. For this reason, I have argued elsewhere that this period might be viewed as the genesis of a second Axial Age. See Keiron Le Grice, *The Archetypal Cosmos: Rediscovering the Gods in Myth, Science and Astrology* (Edinburgh: Floris Books, 2010), 292–294.

3. Tarnas, *Cosmos and Psyche*, 418. I also discuss the correlations between the Neptune-Pluto conjunction and the evolution of consciousness in *The Archetypal Cosmos*.

4. Tarnas, *Cosmos and Psyche*, 418.

5. A point made by Alan Watts citing G. R. Taylor. See Alan Watts, *The Two Hands of God: The Myths of Polarity* (1963; Repr. London: Rider & Company, 1978), 39–40.

6. Watts, *Two Hands of God*, 41.

7. These themes are addressed by Nietzsche in *On the Genealogy of Morals, Beyond Good and Evil*, and elsewhere; and by Jung in *Answer to Job, Aion,* and in a number of other places throughout his collected works.

8. See, for example, Friedrich Nietzsche, *Thus Spoke Zarathustra*, translated by R. J. Hollingdale. (London: Penguin, 1969).

9. For a discussion of diachronic sequencing of archetypal patterns, see *Cosmos and Psyche*, 149–158.

10. Carl Gustav Jung, *Memories, Dreams, Reflections*, edited by Aniele Jaffe, translated by Richard Wilson and Clara Wilson (1963; Repr. New York: Vintage Books, 1989), 26–27.

    The Moon is the archetypal principle most closely associated with childhood; Neptune is the principle associated with dreams and visions; and Uranus with awakening, revelations, sudden insights, and jolting experiences. Occurring in dynamic alignment in Jung's chart, these three principles together find expression as the sudden revelatory spiritual awakenings that defined Jung's childhood and anticipated the future direction of his life and work. Yet if the Moon-Uranus and Sun-Neptune combinations refer to the archetypal conditions behind the occurrence of such dreams, the actual content of the dreams unmistakably reflects themes associated with the Moon-Saturn-Pluto complex. In general terms, the Moon-Pluto-Saturn complex sometimes symbolizes childhood experiences of a dark, disturbing, traumatic, and fateful character. It was in keeping with the nature of the Sun-Neptune archetypal complex, that for Jung these experiences took place in the world of dream, vision, and fantasy, and not primarily as external events.

11. Jung, *Memories, Dreams, Reflections*, 56.

12. Jung, *Memories, Dreams, Reflections*, 21–39.

13. Jung, *Memories, Dreams, Reflections*, 202.

14. Jung, *Memories, Dreams, Reflections*, 56.

15. For his discussion of Apollonian and Dionysian sensibilities in Greek tragedy, see Nietzsche's "The Birth of Tragedy" in *Basic Writings of Friedrich Nietzsche*, translated by Walter Kaufmann (New York: Modern Library, 2000).

16. Plutonic themes in the human imagination are evident in the arts, especially in the Primitivist movement in painting, with its focus on a return to a primitive natural life, in Picasso's incorporation of African influences into his art (1907–1909), and in Fauvism (*les fauves* means "wild beasts") roughly between 1900 and 1910.

17. Jung, *Memories, Dreams, Reflections*, 224.

18. Jung, *Memories, Dreams, Reflections*, 203.

19. Stephen Arroyo, *Astrology, Karma, and Transformation: The Inner Dimensions of the Birth Chart* (Sebastopol, CA: CRCS, 1978).

20. Jung, *Memories, Dreams, Reflections*, 201.

21. Jung, *Memories, Dreams, Reflections*, 201.

22. For the contrast between Uranus-Pluto and Saturn-Pluto see *Cosmos and Psyche* chapters IV and V. The content of Jung's fantasy images not only reflected the Moon-Saturn-Pluto complex in his natal chart but also possessed a wider collective significance in that his fantasies conveyed themes associated with the Saturn-Pluto world transit of 1912–1916. This close attunement to collective events was a recur-

ring pattern in Jung's experience. In particular, surveying Jung's life, one is struck by the influence on his psyche of both world wars, the first coinciding with his confrontation with the unconscious, the second marking the beginning of his comprehensive study of alchemy. See Deirdre Bair, *Jung: A Biography* (Boston, MA: Little, Brown and Company, 2003), chapter 30 "Rooted in Our Soil"). The Plutonic energy that had impressed itself on the culture through the imagination, myth, spirituality, and dream during the Neptune-Pluto conjunction became manifest more concretely during the two world wars, both beginning under hard aspects of Saturn and Pluto. Through his inner fantasies, Jung directly experienced the pressure and sense of dark foreboding manifesting across Europe. During his "confrontation with the unconscious" and again in the 1930s when he reflected on the Wotan possession occurring in Nazi Germany, Jung became particularly attuned to the darker undercurrents lurking beneath the facade of rational self-determination, morality, and civilized culture. "We know nothing of man," Jung bemoaned in an interview with the BBC's John Freeman. The greatest danger to the world, Jung realized, is our collective ignorance of the long-neglected powers of the unconscious.

23. Carl Gustav Jung, *Mysterium Coniunctionis*, 2nd edition, volume 14 of *The Collected Works of C. G. Jung*, translated by R. F. C. Hull (1955–1956, 1970; Repr. Princeton: Princeton University Press, 1989), 272.

24. Jung, *Memories, Dreams, Reflections*, 200–201.

25. Richard Tarnas has made this observation with respect to Jung's natal Sun-Neptune square alignment.

26. Jung, *Memories, Dreams, Reflections*, 194. This episode in Jung's life, it is worth noting, took place under world transits between Uranus and Neptune (a conjunction, 1899–1918) and Saturn and Pluto (a conjunction, 1913–1916), coinciding with the outbreak of the First World War.

27. Jung, *Memories, Dreams, Reflections*, 202.

28. Jung, *Memories, Dreams, Reflections*, 225.

29. Carl Gustav Jung, *The Red Book*, edited by Sonu Shamdasani, translated by Mark Kyburz, John Peck, and Sonu Shamdasani (New York: W. W. Norton & Co., 2009), 370.

30. Jung, *Red Book*, 370.

31. Jung, *Red Book*, 232.

32. Jung, *Red Book*, 232.

33. Jung, *Red Book*, 232.

34. We can see examples of this increasing sense of perspective coinciding with the start of the Neptune-Pluto sextile in works such as Erich Neumann's *The Origins and History of Consciousness* (1954) and Joseph Campbell's *The Hero With a Thousand Faces* (1949), both of which apply the theories of Freud and especially Jung to illuminate mythology, spiritual transformation, and the evolution of consciousness. As Jung notes in his introduction to Neumann's book, developing a coherent vision of archetypal psychology was impossible in the early stages of the movement when ideas were initially bursting forth. By the time of the Neptune-Pluto sextile, however, the insights of depth psychology were more established, and could be viewed as a whole, interpreted, and contextualized.

35. Jung's discovery of alchemy through Richard Wilhelm's translation of *The Secret of the Golden Flower*, enabled Jung to abandon work on *The Red Book* from 1930 onwards. In alchemy, he felt he had discovered an objectively existing model of the transformative process he had been through. See Jung, *Red Book*, 360.

36. Jung, *Mysterium Coniunctionis*, 359.

37. Jung, *Mysterium Coniunctionis*, 41.

38. Jung, *Mysterium Coniunctionis*, 310.

39. Jung, *Mysterium Coniunctionis*, 472.

40. Jung, *Mysterium Coniunctionis*, 472. For Jung's treatment of this topic, see his discussion of the stages of the conjunction, especially those passages relating to the attainment of the *unio mentalis*, in the final section of *Mysterium Coniunctionis*, 457–553.

41. Jung, *Mysterium Coniunctionis*, 225.

42. Jung, *Mysterium Coniunctionis*, 25.

43. Jung, *Mysterium Coniunctionis*, 95.

44. Jung, *Mysterium Coniunctionis*, 14.

45. Jung, *Mysterium Coniunctionis*, 196.

46. Jung, *Mysterium Coniunctionis*, 204

47. Jung, *Mysterium Coniunctionis*, 117.

48. Jung, *Mysterium Coniunctionis*, 112.

49. Jung, *Mysterium Coniunctionis*, 122, 114.

50. Jung, *Mysterium Coniunctionis*, 128.

51. Jung, *Mysterium Coniunctionis*, 114. For Edward Edinger's discussion of this topic, see *Ego and Archetype: Individuation and the Religious Function of the Psyche* (Boston, MA: Shambhala, 1972), 92–93. Edinger quotes Rivkah Scharf Kluger's description of Satan as "truly Lucifer, the bringer of light. He brings man the knowledge of God but through the suffering he inflicts on him, Satan is the misery of the world which alone drives man inward, into the 'other world'" (*Ego and Archetype*, 93).

52. Jung, *Mysterium Coniunctionis*, 144–145.

53. Jung, *Mysterium Coniunctionis*, 142.

54. Jung, *Mysterium Coniunctionis*, 489. This process is also referred to as the *distractio* which, according to Jung, brings about the dissolution of the state in which "the affectivity of the body has a disturbing influence on the rationality of the mind" (*Mysterium Coniunctionis*, 471).

55. Jung, *Mysterium Coniunctionis*, 473, 488 489.

56. Jung, *Mysterium Coniunctionis*, 519–520.

57. Carl Gustav Jung, *Aion: Researches into the Phenomenology of the* Self, volume 9, part I of *The Collected Works of C.G. Jung*, translated by R. F. C. Hull (London: Routledge, 1951), 64.

58. Jung, *Mysterium Coniunctionis*, 89.

59. Jung, *Mysterium Coniunctionis*, 521.

60. Jung, *Mysterium Coniunctionis*, 534.

61. Jung, *Mysterium Coniunctionis*, 490.

62. For more on this topic, see Le Grice, *Archetypal Cosmos*, 44–54, 272–275, 282–287, 289–300.

Bibliography

Arroyo, Stephen. *Astrology, Karma, and Transformation: The Inner Dimensions of the Birth Chart.* Sebastopol, CA: CRCS, 1978.

Bair, Deirdre. *Jung: A Biography.* Boston, MA: Little, Brown and Company, 2003.

Campbell, Joseph. *The Hero with a Thousand Faces.* 1949. Repr. London: Fontana, 1993.

Edinger, Edward. *Anatomy of the Psyche: Alchemical Symbolism in Psychotherapy.* Peru, IL: Open Court, 1994.

Jung, Carl Gustav. *The Collected Works of C. G Jung.* 19 vols. Trans. R.F.C. Hull Princeton: Princeton University Press and London: Routledge & Kegan Paul, 1953–1979.

——. *Aion: Researches into the Phenomenology of the Self.* Volume 9, part I of *The Collected Works of C.G. Jung.* Translated by R. F. C. Hull. London: Routledge, 1951.

——. *Answer to Job. The Problem of Evil: Its Psychological and Religious Origins.* 1960. Translated by R. F. C. Hull. Repr. Cleveland, OH: Meridian, 1970.

——. *Memories, Dreams, Reflections.* 1963. Edited by Aniele Jaffe. Translated by Richard Wilson and Clara Wilson. Repr. New York: Vintage Books, 1989.

——. *Mysterium Coniunctionis.* 2nd Edition. Volume 14 of *The Collected Works of C. G. Jung.* 1955–1956, 1970. Translated by R. F. C. Hull. Repr. Princeton: Princeton University Press, 1989.

——. *Psychology and Alchemy.* 2nd edition. Volume 12 of *The Collected Works of C. G. Jung.* Translated by R. F. C. Hull. Princeton: Princeton University Press, 1968.

———. *The Red Book.* Edited and Introduced by Sonu Shamdasani. Translated by Mark Kyburz, John Peck, and Sonu Shamdasani. New York: W. W. Norton & Co., 2009.

———. "The Spiritual Problem of Modern Man." 1928/1931. In *Civilization in Transition.* 2nd Edition. Volume 10 of *The Collected Works of C. G. Jung.* Translated by R. F. C. Hull. Princeton: Princeton University Press, 1989.

Jung, Carl Gustav, and John Freeman. *Face to Face Interview: Professor Jung.* London: BBC Television, 1959.

Le Grice, Keiron. *The Archetypal Cosmos: Rediscovering the Gods in Myth, Science and Astrology.* Edinburgh: Floris Books, 2010.

Neumann, Erich. *The Origins and History of Consciousness.* 1954. Repr. Princeton: Princeton University Press, 1973.

Nietzsche, Friedrich. *Basic Writings of Nietzsche.* Translated by Walter Kaufmann. New York: Modern Library, 2000.

Nietzsche, Friedrich. *Thus Spoke Zarathustra.* Translated by R. J. Hollingdale. London: Penguin, 1969.

Tarnas, Richard *Cosmos and Psyche: Intimations of a New World View.* New York: Viking, 2006.

Watts, Alan. *The Two Hands of God: The Myths of Polarity.* 1963. Repr. London: Rider & Company, 1978.

Richard Feynman

An Archetypal Analysis of No Ordinary Genius

Clara Lindstrom

As a teenager, I found Richard Feynman absolutely irresistible. Someone gave me a copy of his book *Surely You're Joking, Mr. Feynman!* when I was in high school, and reading it was a visceral pleasure. My whole being seemed to resonate like a tuning fork as I devoured the first-person accounts of his exploits. I not only looked up to him, I wanted to *be* him, to live life the way he lived it. A character whose brilliance was matched only by his rabid pursuit of mischief and adventure, Feynman played his way through life. He followed what he loved, never submitting to others' expectations of what a respectable scientist should be; on the contrary, he assiduously avoided confining definitions at all costs. A Nobel Prize winning physicist who played in a Brazilian samba band and learned to paint nudes; a prank-playing junior scientist on the Manhattan Project who cracked safes (filled with nuclear secrets) in his spare time for kicks; a respected California Institute of Technology professor who prepared his lecture notes in a strip club—this was a man who *loved life*, and he lived it with extraordinary audacity and originality.

Driven by an inexhaustible curiosity and a desire to know how things worked, he seemed to have been not so much birthed as shot out of a cannon, a fiery ball of energy fueled by an insatiable *"why . . . ?"* His "why" was not satisfied by superficial answers or rote explanation, either. From his childhood days spent taking apart and reassembling radios to his later fame as a member of the U.S. Presidential Commission investigating the Space Shuttle explosion, the man was obsessed with the pleasure of finding things out. He was forever burrowing down to the hidden roots of problems or pulling back the proverbial curtain to expose what or who was operating behind the scenes, usually to the consternation of more conventional types around him. Given that I was an ambitious, straight-A, hyper-inquisitive teenager figuring out how I wanted to be in the world, he seemed the most dazzling definition of success I had ever seen.

Figure 1 **Birth Chart of Richard Feynman**

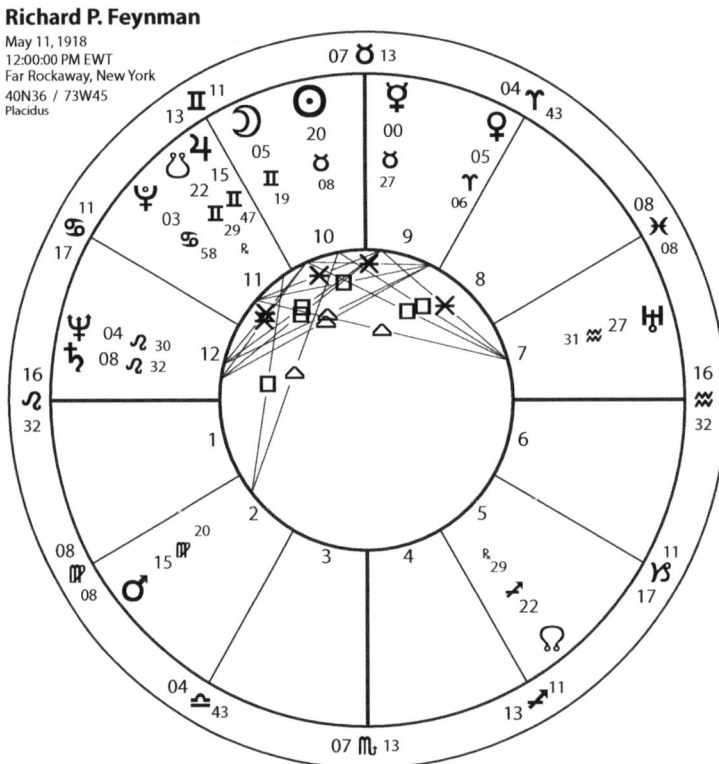

Richard P. Feynman
May 11, 1918
12:00:00 PM EWT
Far Rockaway, New York
40N36 / 73W45
Placidus

My starry-eyed adulation has tempered a bit since then, but, given my enduring affection for Feynman, this will not be a soberly balanced portrait of the man; indeed, I deliberately chose someone with whom I have some resonance. I will include a brief description of each planetary archetype with its introduction here, but my assumption is that the reader is already familiar with the panoply of archetypal characteristics associated with each of the planets. Although addressing every aspect of Feynman's chart is not possible within the scope of this paper, I will examine how particular planetary configurations exhibit a spectrum of characteristics, revealing how each represents a complex interaction of multivalent archetypes. His Saturn-Neptune conjunction—about which an entire thesis could be written—is a particularly rich and fruitful aspect of his chart on which I will focus particularly, and upon which all his other complexes may be seen to rest. Having Saturn and Neptune in aspect myself, I will offer a personal perspective on the nature of this dynamic and suggest that my analysis might be used as a lens through which to view Feynman's actions as well.

Astrological Analysis

What is it that stands out about Richard P. Feynman? What sort of traits did he possess that set him apart from others who I admire, but who do not elicit this sort of wriggling glee when I think about them? First and foremost, he had the aforementioned Saturn-Neptune conjunction (just over four degrees from exact), which immediately puts him in my camp, so to speak, given my own Saturn-Neptune opposition (five degrees). Feynman was, by all accounts, a scientific genius. But more pointedly, he approached life with the skeptical, probing, suspicious-of-consensual-reality manner of one for whom the tension between the Saturn and Neptune principles is always present. Because Saturn represents the archetypal principle of structure, contraction, materiality, and boundaries, and Neptune symbolizes flow, dissolution, transcendence (of corporeal reality), and the tendency to merge, these two archetypes may be viewed, in a certain sense, as opposite ends of an energetic spectrum. Where the Saturn principle represents crystallization, the Neptune principle represents dissolution; where Saturn brings limits, duty, restraint, and judgment, Neptune symbolizes formlessness, imagination, surrender, and mystical union. In a lecture Feynman gave at the Esalen Institute in California, he inadvertently captures the tense yet creative interplay between these two archetypes:

> The game I play is a very interesting one. It's imagination [Neptune] . . . in a tight straight jacket [Saturn], which is this: that it has to agree with the known laws of physics . . . it requires imagination to think of what's possible, then it requires an analysis back, a

checking in, to see whether it fits, it's *allowed*, according to what is known.[1]

Brought up Jewish, he was an avowed atheist by the time he was in high school, having discovered that the laws of nature did not support any of the miracles he had been told about in scripture.[2] In a hilarious yet poignant example of this Saturn-Neptune complex, he tells the story of how, at age eleven, he had been listening for years to various Sunday-school accounts of the Jewish people and the miracles that had saved them throughout their history. Upon questioning the rabbi more closely about one character, however, he was shocked when the rabbi told him that this person wasn't a real individual, but rather a composite character meant to represent the suffering of the Jews more personally. As he tells it:

> That was too much for me. I felt terribly deceived: I wanted the straight story—not fixed up by somebody else—so I could decide for myself what it meant. But it was difficult for me to argue with adults. All I could do was get tears in my eyes. I started to cry, I was so upset. (The rabbi) said "what's the matter?" I tried to explain. "I've been listening to all these stories, and now I don't know, of all the things you told me, which were true, and which were not true! I don't know what to do with everything that I've learned!" I was trying to explain that I was losing *everything* at the moment because I was no longer sure of the data, so to speak. Here I had been struggling to understand all these miracles, and now . . . well, (this admission) solved a lot of miracles, all right! But I was unhappy.[3]

That feeling of having the ground shift underneath one, of not knowing what can be counted on as real, of trying desperately to pin things down (Saturn) but experiencing reality as morphing fluidly (Neptune), is a classic example of the Saturn-Neptune complex. Finding oneself in the midst of this archetypal interplay can be excruciating. This illustrates why, when he found out that Santa Claus isn't real, he wasn't upset but rather "relieved that there was a much simpler phenomenon to explain how so many children all over the world got presents on the same night!"[4]

A more nuanced expression of the Saturn-Neptune complex, and again something with which I resonate profoundly, is the rejection of an external code of ethics governed by religion or social convention in favor of an immanent, vividly experienced internal morality. The result of the Neptunian archetype, which is associated with the ideal and the spiritual, interacting with the Saturnian qualities of established order, rules, and conventions, this manifestation of the complex is particularly layered and intricate. Subjectively, this feels less like a mere socially sanctioned idea of what is right or wrong and more like a profound need to originate all one's thought and action from a place of *total authenticity*. To do otherwise is exceedingly uncomfortable. Viewed from this angle, it can be seen as the Saturnian reality principle testing and piercing the Neptunian façade and illusion of consensus reality, i.e., one perceives that others are being taken-in by the delusion and escapism (Neptunian shadow qualities) of mass culture and one responds with a heightened focus on judgment, discernment, and discipline (Saturnian characteristics).

The constant tension between the Saturn and Neptune principles creates a sort of metaphorical friction that shears off the rough edges of lazy acceptance or complacence in an individual. The shifting fluidity of Neptune's multiple realities in relationship to Saturn's insistence upon something fixed and immutable—something that can be counted on—forces the individual to eschew popular wisdom and find some point *within* that is real, solid, and true. This does not happen once, but over and over throughout one's life in a consistent practice of refinement. In many ways this is reminiscent of the process that takes place during contemplative prayer, or a dark night of the soul, when the competing voices of outside authorities must be stilled in order to find internal solid ground, an existential perch of sorts. Everything else may be in question, but that touchstone of truth inside is unwavering. And from this initial plank, a stable foundation can gradually be built. (I would also contend that this pared-down place within is what connects us directly to universal spiritual laws, i.e., to a universal ideal of Truth, or to God, but this is my experience of Saturn-Neptune and not Feynman's.)

Because each plank added to this "scaffolding of truth" is so hard-won, any major deviation from one's internal authenticity can be quite destabilizing. Examples of this might include not voicing one's real opinion at work because it would jeopardize one's job, joining in racist or misogynistic jokes to fit into a group, or simply telling one's beloved "everything is fine" when that is in fact far from the case. This may be felt not so much as an inconsequential straying from one's moral compass but more like a splitting off from the *only solid reality that one knows*: suddenly the ground is shifting crazily again. Intense anxiety or a kind of "existential dread" typically accompanies this, and the agonizing discomfort virtually forces one back to one's center—i.e., towards words or action that express the authentic self—thus alleviating the psychological torment.

Feynman, whose first wife died within just a few years of their marriage, aptly illustrates this in a passage about breaking the news of her terminal illness. In those days (the 1940s) it was thought best not to alarm the patient with dire prognoses. Although Feynman and his beloved had made a pact of absolute honesty and he wanted to tell her the truth (a dynamic underpinned by Venus in a trine aspect to his Saturn-Neptune), he was virtually forced to lie to her:

> At home, everybody worked on me: my parents, my two aunts, our family doctor; they were all on me, saying I'm a very foolish young man who doesn't realize what pain he's going to bring to this won-derful girl by telling her she has a fatal disease. "How can you do such a terrible thing" they asked, in horror. "Because we have made a pact that we must speak honestly with each other and look at everything directly. There's no use fooling around. She's gonna ask me what she's got, and I cannot lie to her!" "Oh, that's childish" they said – blah, blah, blah. Everybody kept working on me, and said I was wrong. I thought I was definitely right…that telling her the truth was the right way to handle it. But finally . . . I couldn't take it anymore. . . . I go to the hospital to see Arlene—having made this decision—and there she is, sitting up in bed, surrounded by her parents, somewhat distraught. When she sees me, her face lights up…nodding at her parents, she continues "they're telling me I have glandular fever, and I'm not sure whether I believe them or not. Tell me, Richard, do I have Hodgkin's disease or glandular fever?" You have glandular fever, I said, and I died inside. It was terrible—just terrible![5]

The above exchange hits home all the more when one considers that Feynman had Mercury squaring his Saturn-Neptune conjunction (eight degrees and four degrees from exact, respectively). Mercury is the planet associated with the archetype of thought, speaking, communication, the ability to understand and articulate, and the impulse to connect and mediate.[6] With Mercury in the aforementioned position, he was compelled to outwardly convey his Saturn-Neptune commitment to authenticity; unable to sustain the lie, he told his wife the truth a week later, which was a relief to them both. One could also argue, here, that it was a shadow aspect of Neptune—the potential to become merged or overwhelmed in the sea of values or ideals of one's milieu—that took him initially off balance, or away from his own commitment to truth. In this case, the Saturn principle (defining one's actions as separate and independent) resurfaced a week later and served as a perfect foil and corrective.

I will now address something that would seem to be in total opposition to the way I have explained Feynman's Saturn-Neptune complex above, i.e., in terms of constantly looking for internal solid-ground. The following observations simply reveal, however, a different facet of the same fundamental principle, which is that Saturn in hard aspect to Neptune engenders a *singularly acute relationship to uncertainty.* For ironically, the same Saturn-Neptune complex that impels one to valiantly adhere to an internal truth is *also* one that allows an individual to resign himself to *not knowing anything for sure.* How can this be? It seems an odd paradox, but one that testifies to the multivalent expression of the archetypes: one profound well can feed many surface springs.

The same man who agonized over lying to his wife and stubbornly insisted on speaking his truth even when the stakes were extremely high because to do otherwise would be a betrayal of reality, of sorts, is also the man who declared:

> I can live with doubt and uncertainty and not knowing. I think it's much more interesting to live not knowing than to have answers which might be wrong. I have approximate answers and possible beliefs and different degrees of certainty about different things, but I'm not absolutely sure of anything . . . I don't feel frightened by not knowing things, by being lost in a mysterious universe without having any purpose, which is the way it really is so far as I can tell. It doesn't frighten me.[7]

How does one person inhabit two such seemingly incongruent positions? Attempts at explanation bring one up against the difficulty of articulating an archetypal feeling that lives more vividly as a somatic sensation or picture in the mind than as words strung together on a page. Nevertheless, my sense is that the crux of the matter lies, to an extent, in the perceived distinction between outer and inner; collective and personal. The consensual world "out there" is in constant question—nothing that anyone else says can be taken as a given *unless* it can be held up against one's personal experience and found to be resonant. Thus the litmus test for any statement of fact is this checking-in with one's own meticulously gathered findings (in the case of scientific experiment) or deeply considered feelings (in the case of moral decisions). It is a continual dance between a Neptunian sea of what *might be* versus a Saturnian pin-point of *what is*.

This could be imagined as a person floating on a small but solidly built raft (Saturn) in an unimaginably vast, ever-undulating ocean (Neptune). In order to survive, this person must repeatedly dive into the swirling, teeming, life-giving waters in search of new ideas and nourishment but, once some tasty morsel has been caught or new information gathered, she must return to the raft in order to lay it out on a dry surface, get a good look at it from all sides, hold it in her hands, test its weight, perhaps take a bite out of it, and only then decide for herself whether or not it is good. Because if it *isn't* good—if it turns out not to taste quite right or appears to be somehow at odds with the rest of the carefully chosen pieces on the vessel—it will be unceremoniously tossed off the tiny raft and back into the sea.

So this person is always in motion, always diving. And it is in fact the sea of uncertainty that supports and keeps the little raft afloat, so she may be quite comfortable with this arrangement, surrounded by not-knowing on all sides—even joyfully swimming in it—as long as she can return to the raft. *Always* there is a returning to the raft, which may itself have changed position a bit during one of these forays, as the seas naturally ebb and flow. But it is unequivocally the same raft, and serves as the personal proving ground for any amorphous blob that may be plucked from the water.

This notion of fluidity in constant and close relationship to a fixed bit of hard matter handily describes, moreover, the nature of Richard Feynman's chosen profession. That he was a skeptical scientist is already in line with the thought experiment; that he was in the field of quantum physics, *specifically*, is the proverbial icing on the cake. To excel in this field it helps to be unusually comfortable with the uncertainty—some might say mystery—of the universe. No other discipline so forthrightly and manifestly addresses the astonishing nature of physical reality: the wonderful fact that energy behaves both as waves (Neptune) *and* as particles (Saturn). This is the beating heart of the Saturn-Neptune complex, writ large; this is the constant tension between two apparently opposing versions of reality that one is obliged to hold. Feynman openly addressed this essential tenet of both the Saturn-Neptune complex and quantum physics—the issues of paradox and uncertainty—repeatedly. He exhorted his students to relax about not knowing; to enjoy the game of searching. It was a major and recurrent theme in his life, spawning lectures, discussions, and even a book entitled *The Meaning of It All*, which contains the chapters "The Uncertainty of Science" and "The Uncertainty of Values." In his own words he explains:

> The scientist has a lot of experience with ignorance and doubt and uncertainty, and this experience is of very great importance, I think . . . scientific knowledge is a body of statements of varying degrees of certainty—some most unsure, some nearly sure, but none *absolutely* certain. Now, we scientists are used to this, and we take it for granted that it is perfectly consistent to be unsure, that it is possible to live and *not* know. But I don't know whether everyone realizes this is true.[8]

A further expression of the Saturn-Neptune complex—and the perfect flip side of Feynman's intimate experience with uncertainty and *not* knowing—was his vehement position on the difference between learning something superficially and pretending to know it versus really *knowing and understanding* it. Throughout his life he would repeat to his students a lesson he had learned early on from his father, namely: if you have merely memorized something you do not know the first thing about it (this would be akin to abandoning your own raft, jumping onto an ocean liner, and relying on its navigation systems to guide you; i.e., you have learned nothing about how to navigate the world for yourself). He was profoundly dismayed when, while teaching in Brazil, he realized that his students were taking notes and carefully memorizing everything he said but that, when asked to give actual examples of the phenomena they were supposedly learning, they were completely stymied. They had memorized the information but, for the most part, had no idea what it meant or how it applied in the real world! While appearing to honor Saturnian ideals of diligence, judgment, and adherence to empirical reality they were in fact creating an illusory fiction (one shadow side of the Neptune archetype) that merely projected the *outward appearance* of scientific rigor. Worse, their behavior could be seen as sacrificing the *positive* Neptunian ideals necessary for Feynman's brand of science—imagination, dissolution of strict notions of material reality, and an immersion in the realm of the ideal or intangible—for the sake of maintaining this Saturnian façade.

Feynman ruminated about this state of affairs all year, obsessing over it in much the same way that someone might after having been personally wronged. He tried in vain to get the students to ask questions, use their imaginations, and experiment with real materials (the Saturn archetype in evidence here, representing the reality principle and the practical application of knowledge). At the end of his year in Brazil, he delivered a shockingly frank lecture to the university students, faculty, and several government officials pointing out that the system for teaching science in Brazil was a complete charade, saying that he "couldn't see how anyone could be educated by this self-propagating system in which people pass exams, and teach others to pass exams, but nobody *knows* anything."[9] His hope was that by forcing people to listen to his exposé he would oblige them to see what he so clearly saw, thereby instigating change for the better (he did, to an extent).

In this respect his behavior was, again, a prime example of one classic way in which the opposing archetypal principles associated with Saturn and Neptune intricately intertwine and co-manifest: a propensity to unmask hypocrisy, expose social injustice, and continually call into question prevailing wisdom.[10] My assertion is that this inclination may be viewed, in instances like Feynman's, as an outgrowth of the Saturn-Neptune individual's proclivity to feel worldly injustice on a deeply personal level. I am using the term "injustice" in a broad sense here, meaning it to encompass in fact any circumstance that exists due to what the Saturn-Neptune individual perceives (often quite rightly) as the misguided action, ignorance, or naiveté of others. This could describe anything from racial discrimination or misuse of religious authority to fatuous political promises or the poor administration of an organization, as Feynman observed in the case of the Brazilian university system.

When the goings-on around such a person do not match up to his personal standards (assiduously tested on the proving-ground of the Saturnian raft) it may feel not only like a travesty of justice "out there in the world" but like a barb that pierces the heart of his own exhaustively considered moral order. It is painful. Because he has invested so much in testing the rules of reality and morality, he is confident they are correct, or are, at least, the best approximation available. Having spent his life navigating the tensions between the diametrically opposed archetypal realms associated with Neptune and Saturn, he assumes that his findings are true. They are an accurate description of reality. In a certain sense, his very existence depends upon it, for they are what keep him afloat. As such, he is totally identified with his convictions. He naturally expects this reality to extend out in all directions and provide a firm foundation for everyone to stand on. When the rest of the world does not in fact jump onboard, he is genuinely mystified as to why, and feels it acutely.

Of course, the person who experiences the Saturn-Neptune archetypal complex in this way may be mostly unconscious of the dynamics described here. Rather than think "this is *my* conception of reality or morality and everyone should get onboard," he most likely assumes "I am simply seeing reality (or morality) *as it is*; why is everyone else acting insanely, immorally, or contrary to the laws of nature?" His views are the lens through which he sees, and may therefore be invisible to him. Granted, most people probably take their own views about reality to be the ontological facts governing reality, but the individual deeply attuned to the Saturn-Neptune complex perhaps experiences this in a keener sense. Whereas the former might feel "there are forces out there in the world acting in wrong-headed ways and I don't agree," the latter might feel that "there are forces out there in the world acting in wrong-headed ways and it feels personally painful or injurious." The suffering of others "out there" may become quite literally his own suffering. Due to this almost physical uneasiness, such a Saturn-Neptune individual may feel compelled, as Feynman did, to take some kind of action—not only to ameliorate his discomfort and right a wrong, but in an effort to *have the world make sense again;* to nudge the external world into alignment with his powerfully present internal world and thus help create reality as it is *supposed to be*.

Granted, it is important to remember that due to the multivalent nature of the planetary archetypes, any description of one particular way in which a complex manifests will never apply to *all* individuals whose charts include this same alignment. There are most certainly individuals with Saturn and Neptune in alignment who do not share this same sense of things. Nevertheless, Feynman's indomitable impulse to unmask fakers, expose deceivers, and correct the ways of wrong-doers, i.e., people he judged to be acting, whether intentionally or not, in opposition to the Moral Order (such as the students and teachers in Brazil), is one powerful and quintessential manifestation of the Saturn-Neptune dialectic.[11]

As we have seen, the potent interaction of Saturn-Neptune archetypal forces figured prominently in Feynman's life. Much as it provided a framework for his way of being in the world, however, his Saturn-Neptune complex did not exist in isolation. As I mentioned, he enjoyed the good fortune (I would assert) of having Mercury squaring his Saturn-Neptune conjunction. The archetype of thought, intellect, communication, learning, and education, Mercury in this alignment figures prominently in his concerns about the Brazilian educational system and the sharp lecture he delivered at the end of the year. It also sheds light on one of the reasons for Feynman's immense popularity as a teacher and lecturer. Genius though he was, he refused to dress things up in technical jargon and made it a point to talk about science in a way that others could understand. Indeed, he was often called the "Great Explainer," and "gained a reputation for taking great care (Saturn) when giving explanations (Mercury) to his students and for making it a moral duty (Saturn-Neptune) to make the topic accessible. His guiding principle was that if a topic could not be explained in a freshman lecture, it was not yet fully understood."[12] Viewed from one angle, there could not be a more straightforward example of his Saturn-Neptune conjunction, squared by Mercury. (He was not always successful, of course, as his *Feynman Lectures on Physics* were meant for freshmen but were in fact more suited to his colleagues. Viewed from another angle, in fact, the Mercury-Saturn complex can show up as a tendency to be impenetrable and hard to understand; the Mercury-Neptune complex vague or confused; so it is to his credit that he was typically able to explain intricate subjects so well.)

Moving beyond Neptune's involvement to look at just his Mercury square Saturn, we see a host of associated traits that aptly describe the professor: self-critical intellectual rigor, practical thinking applied to the real world, an unusual clarity and simplicity of expression, close observation and attention to detail, and—perhaps most strikingly—a mental stubbornness and tenacity when solving seemingly intractable problems. With regard to the first characteristic, Feynman was constantly calling himself a "dope" and an "idiot" and going out of his way to describe situations in which he was confused or didn't catch onto things (the harsh judgment of the Saturn principle against his Mercury-Neptune complex). He berated himself for not properly diagnosing his wife's illness, for example, although he was just twenty-four at the time and not a doctor. Certain colleagues considered his self-deprecating tendencies a false modesty designed to get attention and resented him for it, and it is also true that, often as not, Feynman relished telling anecdotes in which he came out the mental victor. So he clearly knew he was smart (Mercury), but like most exceptionally sharp people, he was often the first to see his own intellectual failings (Saturn). This natural competitiveness and sensitivity to victory and defeat also points to his Mars trine Sun, strengthened further by his identification as the brilliant Sun-Uranus, brightly shining individual.

His Mercury-Saturn clarity of expression also evinced itself far beyond his professional field, for it practically leaps off the pages of his several books. No one could tell a story like Richard Feynman. Even with his notorious disregard for proper grammar and conventions of speech (Uranus sextile Mercury), he managed to get right to the heart of something without beating around the bush, and he communicated in such a way that the listener could *feel* his intended point more than just hear it. Understanding others and being understood in a plain-spoken kind of way was critically important to him (Neptune influencing Saturn square Mercury, as it relates to his no-*façade* self expression, and Neptune-Mercury as the empathetic, sensitive communicator) and he told multiple stories about inadvertently befuddling certain fancy types with his frank speech when vague platitudes and social niceties would have been more appropriate (this also brings in his Pluto-Mercury-Uranus alignment, which I will address below).

Finally, his Mercury-square-Saturn ability to see a problem through to its solution was legendary, and in large part what took him so far in his career. In reference to fixing the stereo of one of his mother's friends, he wrote

> I finally fixed it because I had, and I still have, persistence. Once I get on a puzzle, I can't get off. If my mother's friend had said 'Never mind, it's too much work,' I'd have blown my top, because I want to beat this damn thing, as long as I've gone this far. I can't just leave it after I've found out so much about it. I have to keep going to find out ultimately what is the matter with it in the end. That's a puzzle drive.[13]

I mentioned above his propensity for disrupting social conventions and regularly creating upheaval when proper behavior was expected. Other than his brilliance in physics, this was perhaps the characteristic for which he was most famous, most loved, and often most reviled. As biographer James Gleick phrased it: "His personality, unencumbered by dignity or decorum, seemed to announce: Here is an unconventional mind."[14] I am one of those people who loves him for this defiance of social convention, and I was hugely gratified to find this major tenet of his personality traceable both to Uranus squaring his Sun (which I will address further on), and to Pluto in a trine to his Uranus with Mercury at the midpoint, in sextile to both. As the archetype of power, primordial energy, compulsion, intensity, and violent upheaval, Pluto seeks cathartic expression, expulsion, destruction, and rebirth; it is nature in the raw. The Uranus archetype is the Promethean impulse: it liberates, rebels, and innovates, bringing revolution, sudden flashes of brilliance, creativity, and paradigm shifts. "With respect to personal character," as Richard Tarnas explains, "Uranus is regarded as signifying the rebel and the innovator, the awakener, the individualist, the dissident, the eccentric, the restless and wayward."[15]

These two planetary archetypes working in concert makes for a potent combination, and there are countless tales Feynman tells when this archetypal complex is vividly evident. One such incident took place at Los Alamos, New Mexico where he was working on the Manhattan Project. A twenty-five-year-old junior scientist, he sat in the back of the room when Niels Bohr, regarded by even the senior figures in physics as a sort of god, arrived for a conference amidst great fawning and reverence. Although Feynman didn't say much, he was surprised the next day by an early morning telephone call from Aage, Bohr's son, who told him that his father would like to meet with him. Shocked, he nevertheless showed up at the lab, where Bohr began to tell him about some ideas he had for making the bomb more efficient. Feynman argues with him:

> I say "No, it's not going to work. It's not efficient . . . blah, blah, blah." So he says, "How about so and so?" I said "That sounds a little bit better, but it's got this damn fool idea in it." This went on for about two hours, going back and forth over lots of ideas, back and forth, arguing. "Well," (Niels) said finally, lighting his pipe, "I guess we can call in the big shots *now*." So then they called in all the other guys and had a discussion with them. (Some time later), his son told me what had happened. The last time he was there, Bohr said to his son "Remember that little fellow in the back over there? He's the only guy who's not afraid of me, and will say when I've got a crazy idea. So . . . we're not going to be able to do it with these guys who say everything is yes, yes, Dr. Bohr. Get that guy and we'll talk with him first." I was always *dumb* in that way. I never knew who I was talking to. I was always worried about the physics. If the idea looked lousy, I said it looked lousy. If it looked good, I said it looked good. Simple proposition. I've always lived that way. It's nice, it's pleasant—if you can do it. I'm lucky in my life that I can do this."[16]

In Feynman's case, the massive power (Pluto) of his intellectual convictions (Mercury, squaring his Saturn-Neptune) enabled him to disregard status and authority in order to break through (Uranus) the scientific pecking order. This same constellation also provided him the fortitude to give his first ever technical presentation, as a graduate student, with Henry Norris Russell and professors von Neumann, Pauli, and Einstein in attendance (it is hard to imagine a more intimidating crowd). Significantly, and as I mentioned before, Uranus is also squaring his Sun (approximately seven degrees from exact), the latter planet representing one's personal identity and self-expression. This complex adds heft and intensity to the above dynamic in a couple of ways: first, he was so heavily self-identified (Sun) with scientific ideas and creative brilliance (Uranus) that it always overshadowed social circumstances, and second, Uranus itself is the archetype of the rebel-trickster, which Feynman personified at all times.

Myriad instances of the Pluto-Uranus-Mercury alignment show up in Feynman's life, and there are more remarkable examples to recount than space permits. Essential to include, however, is a description of how this three-planet complex manifested with a slightly different and more physically literal flavor than in the above account. Feynman talked about how, as an undergraduate at the Massachusetts Institute of Technology (MIT), he had admired the beautifully engineered cyclotron, which was newly built, state of the art, and the size of two whole rooms. A "gold-plated cyclotron," he called it.[17] Arriving at Princeton for his graduate work, the first thing he wanted to do was see *their* cyclotron, so he asked about it and was told to go look in the basement:

> In the *basement*? It was an old building. There was no room in the basement for a cyclotron. I walked down to the end of the hall, went through the door, and in ten seconds I learned why Princeton was right for me . . . in this room there were wires strung *all over the place*! Switches were hanging from the wires, cooling water was dripping from the valves, the room was *full* of stuff, all out in the open. Tables piled with tools were everywhere; it was the most godawful mess you ever saw. The whole cyclotron was there in one room, and it was complete, absolute chaos! I suddenly realized why Princeton (more than elsewhere) was getting results . . . they *built* the instrument; they knew where everything was, they knew how everything worked . . . it was much smaller than the cyclotron at MIT, and "gold-plated"?—it was the exact opposite. When they wanted to fix a vacuum, they'd drip glyptal on it, so there were drops of glyptal on the floor. It was wonderful! [18]

What a picture! Something about this description—the physical descent into the dank, old basement, the dripping water and hanging wires, the exposed guts of the thing—so powerfully brings to mind images of a primordial, Plutonic underworld where an intense intellect (Mercury) is birthing revolutionary, paradigm shifting scientific breakthroughs (Uranus). Indeed, "paradigm shifting" is the operative term here in the sense that Feynman was not merely coming up with new products or useful tools in the tradition of Better Living Through Science, but rather contributing to the field of quantum mechanics, also called "new physics" precisely *because* it was such a departure from Newtonian physics and a radical dissolution (Neptune) of what had heretofore been accepted as the absolute nature of physical reality (Saturn).

It stands to reason that a Sun-square-Uranus kind of character—a Prometheus stealing fire from the heavens—would have the capacity for the kinds of non-conformist leaps in understanding required by this "new physics," and he was described as an iconoclastic and astoundingly original thinker. In the 1993 BBC Horizon program, computer scientist Daniel Hills offered the following observation:

> Scientists in general have a pretty narrow range of things that come up on their radar, that they sort of see, or worry about, or pay attention to, and they tend to reject an awful lot if it doesn't fit into their pattern of things. And...he was just the opposite—he was always *looking* for things that didn't fit into his pattern of things. So he was always deliberately sort of turning any piece of conventional wisdom on its head and questioning it and asking about it.[19]

Of course, all the stories of Richard Feynman would not have been told had he been a shy, retiring hermit type. On the contrary, it was his charisma and fiery personality that helped bring him to the world's attention. Biographer Gleick states "his mystique might have belonged to a gladiator or a champion arm-wrestler" and he possessed a quality that may be described as a forceful, energetic engagement with the world; a distinctive "moving out into the common arena" in order to stir things up, agitate, and potentially get a rise out of others.[20] Viewed from its shadow side, this can come across as a deliberate combativeness or an unwelcome poking at people right where they are weakest. For any bystanders simply wanting ease and harmony, it can be thoroughly exhausting, something like a freight-train of frenzied intensity coming towards one when all one wants to do is curl up in a cozy chair with a book, *not* talk politics or hear an unsolicited lecture on the general theory of relativity.

From another perspective, this quality can be viewed as the irresistible desire to "get into the game"; to passionately engage with others; to act upon the world and create change. "Feisty" would be an understatement. It came as no surprise to me, then, that Feynman has Mars exactly square Jupiter. Associated with the god of war, Mars symbolizes the impulse to act, have impact, move forward and out, defend, be aggressive, competitive, and combative. Jupiter "crowns king" whatever it touches; it expands and enlarges, blesses and honors; it brings abundance, optimism, exuberance, and good fortune. Although the two planets are in hard aspect, the fact that the Mars archetype is interacting with that of Jupiter means that the Niagara Falls of energy can often be felt as more ebullience than belligerence, and may also manifest in service to higher philosophical ideals, principled action, or big picture issues. It bowls one over with its intensity and overconfident assertion, but come a little closer and one might see that its character is more passionate enthusiasm and a desire to share or relate than it is an urge to actually fight. Indeed, from the perspective of the Mars-Jupiter person, I imagine, this kind of engagement can feel like play; it simply feels like how one interacts with people.

Equally relevant, Feynman's Mars is also trine his Sun, which explains both the multiplied intensity and, given the soft aspect, the somewhat graceful expression of his energy: he delighted students with his enthusiastic lectures, was beloved for his declarative statements about how he saw things no matter *what* anyone else thought, and, as I mentioned before, had no trouble asserting himself even in situations where he was the junior fish in the scientific pond.

This Mars-Jupiter-Sun archetypal combination undergirded Feynman's final actions on the Presidential Commission to investigate why the Space Shuttle Challenger exploded. To set the scene, it is necessary to note that he couldn't stand Washington D.C. or anything to do with the Establishment (a reflection of his Uranus-square-Sun rebel nature, and also his Saturn-Neptune aversion to hypocrisy and façade). Almost upon arriving, however, he became deeply involved in political intrigue and relished his role as maverick investigator, sneaking around in classified areas looking for clues that might blow the lid off Pandora's Box and even entitling a chapter in one of his books "gumshoes" (here we see evidence of his Pluto-Uranus-Mercury alignment, with the Pluto-Mercury complex being the quintessential archetype of the detective. Associated with Pluto is the instinct to dig deep and unflinchingly poke around in shadow material, while the Uranus character delights in cutting through traditional decorum and upsetting the status quo. The two of these combined makes for the ideal investigator, and Mercury represents the impetus to communicate the findings). It seemed to him that nobody was digging deep enough and he felt it was his moral obligation to uncover the truth and get an accurate report published (Saturn-Neptune opposite Mercury).

When the time came to write up the group report, he had serious issues with some of what it contained, considering it to be misplaced political pandering in what was supposed to be a sober list of scientifically based recommendations. Meanwhile, he had written his *own* report, which had been so watered down in the process of its integration into the document that it didn't say at all what he had intended. His response to all of this was to issue a statement ordering that his name be removed from the document unless the objectionable bit was taken out and his original contribution was included in full. With his Mars characteristics strengthened by the Jupiter archetype, he continued to defend his position in the hot seat—against NASA and the U.S. government! —deflecting right and left their wheedling attempts to talk him out of it, saying "all of the arguments were like that—none of them was very good, and none of them had any effect. I had thought through carefully what I was doing, so I just stuck to my guns."[21] (This is also an influence of his Saturn-Mercury alignment, Saturn representing the fixed and stalwart adherence to his statement, represented by Mercury).

Likewise, it was his Mars-Jupiter-Sun archetypal complex that came into play when, after letting slip his as-yet unpublished report to two journalists (they were, after all, "two beautiful blondes"—how could he help it?), he called them up to ask that they not use the information. The reply was "we're in the news business Dr. Feynman. The goal of the news business is to get news, and your report is newsworthy. It would be completely against our instincts and practice not to use it."[22] They went head to head late into the night, making calls back and forth to negotiate, the fact that the reporters already had the information putting Feynman in the weaker position. With indomitable Mars confidence, encouraged by Jupiterian optimism, he persisted, and later explained "I was in a very good fettle, for some reason . . . I knew what I needed, so I could focus easily . . . I didn't think there was any law of nature which said I had to give in. I just kept going, and didn't waver at all."[23] The result was, of course, that the women did not print the material.

Feynman's role as maverick scientist bringing truth to the masses was just one aspect of a profoundly multivalent character, however, and a whole array of other traits expressed themselves in Vegas casinos, strip clubs, solitary desert drum sessions, Carnival in Brazil, Swiss brothels, and art studios. Certainly the complexes I have already discussed were present as well in these places. But given his unconventional and rather celebrated behavior with regards to the pursuit of pleasure, appreciation of beauty, and women in general, it stands to reason that he would have a strong Venus in his natal chart, and in fact he does: square Pluto (two degrees) and trine Saturn-Neptune (three degrees, one degree).

The principle of beauty, sensuality, desire, love, and aesthetics, the archetypal Venus brings romance and pleasure, gorgeous adornment, and artistic appreciation. I will touch only briefly on his Venus complex. The limited space I afford it in relation to the others is no indication, however, of its being any less important or intriguing. The following passage is a lovely initial description of one way that his Pluto-Venus-Saturn-Neptune complex manifested, conveyed through Mercury:

> I wanted very much to learn to draw [Mercury], for a reason that I kept to myself: I wanted to convey an emotion I have about the beauty of the world [Venus]. It's difficult to describe, because it's an emotion. It's analogous to the feeling one has in religion [Neptune] that has to do with a god that controls everything in the whole universe . . . how things that appear so different and behave so differently are all run 'behind the scenes" by the same organization, the same physical laws. It's an appreciation of the mathematical beauty [Saturn square Venus] of nature [Pluto], of how she works inside; a realization that the phenomena we see result from the complexity of the inner workings between atoms; a feeling of how dramatic and wonderful it is. It's a feeling of awe —of scientific awe [Saturn-Neptune]—which I felt could be communicated through a drawing [Mercury] to someone who had also had this emotion. It could remind him, for a moment, of this feeling about the glories of the universe.[24]

Richard Feynman, with typical Saturn-Mercury dedication, assigned himself the task of learning to draw in his mid-forties. He became quite accomplished at it and especially loved to draw nude women, eventually being commissioned by the owner of a massage parlor to paint a topless slave girl massaging a brawny Roman soldier—a fitting portrayal of his Venus-Pluto complex. Not surprisingly, this connection was the result of his afternoon visits to the topless bar in his neighborhood, where he would work on physics problems or prepare notes for an upcoming lecture while watching the show. When the restaurant was raided by the police and patrons were asked to testify on its behalf in court, Feynman was one of the very few who took the stand. Everyone else had some kind of reason he couldn't do it: one guy ran a day camp; another owned a business that would lose customers if people knew he came in. Feynman, with his Pluto square Venus, remembered thinking "I'm the only free man in here. I haven't any excuse! I *like* this place, and I'd like to see it continue. I don't see anything wrong with topless dancing. So (he said to the owner) Yes, I'll be glad to testify."[25]

What I find interesting is that, although he was a well known professor at the prestigious California Institute of Technology and undoubtedly the highest profile patron *in* the place, potentially tarnishing his reputation was not a big part of his thinking. I see his actions here as an interaction of Pluto squaring Venus in a trine to Saturn and Neptune: lusty, libidinal enjoyment (Pluto) of pleasure and beauty (Venus) upheld by his internalized, private moral standard (Saturn-Neptune). Of course, it was his Sun square Uranus (the archetypal motif of the "free man") that gave him permission to shock conventional opinion and go public with his convictions, rocking the boat as usual.

Feynman also *loved* Las Vegas—a city that, if archetypes were patron saints, would surely have gaudy altars to Pluto, Venus, and their wanton coupling on every corner. In a palpable way, this city represents one of the more obvious manifestations of the Venus-Pluto complex: an underworld where greed, lust, power, and compulsion (Pluto) bleed into dazzling displays of beauty and sensuality (Venus); a city that comes alive at night, promising twenty-four-hour pleasure and stimulation and reveling in the explosive and cathartic expression of repressed libidinal energies. Understandably, Feynman's delight in the taboo, underground, or socially unacceptable aspects of sensual engagement—typical of someone with Venus in hard aspect to Pluto—enjoyed free reign in Vegas.[26] (I believe there must be a sort of singing resonance that occurs when someone with a prominent archetypal complex in her natal chart finds herself in a locale that embodies those same archetypes, engendering a kind of relief or "at-homeness"). He told with relish various stories of hanging around with show girls (lovely Venus), club owners, and high rollers (powerful Pluto). On the one hand, he repeatedly referred to these types— who many in academic circles would consider seedy characters—as "nice people," talked about how interesting he found them, and lamented that most people cast judgment without knowing the first thing about them.[27]

On the other hand, he could veer towards misogyny and womanizing and was, at the very least, an unrepentant flirt. An entire chapter of his first book is devoted to describing how he learned to pick up show girls by using reverse psychology (in other words, by being an asshole rather than a nice guy). He was tutored by a local player, who counseled him "under no *circumstances* be a gentleman! You must *disrespect* the girls. Furthermore, the very first rule is, don't buy a girl *anything*—not even a package of cigarettes—until you've *asked* her if she'll sleep with you, and you're convinced that she will."[28] His response was

> Well, somebody only has to give me the principle, and I get the idea. All during the next day I built up my psychology differently: I adopted the attitude that those bar girls are all bitches, aren't worth anything, and all they're in there for is to get you to buy them a drink . . . I'm not going to be a gentleman to such worthless bitches, and so on. I learned it till it was automatic.[29]

Although reading this elicits a cringe, it points to how he applied his methodical Saturn-Mercury diligence to diving headlong into murky Pluto-Venus pursuits! He enjoyed the after-hours world, reveled in its intrigue and power dynamics, and as I mentioned above made no secret of his forays into the Plutonic underbelly of sexuality and pleasure. Thus multiple valences of Pluto squaring Venus shimmer back and forth in his chart, from his passionate love and reverence for the intricate workings and beauty of nature to his hedonistic consumption of the Vegas feast.

Conclusion

As is the case with almost anyone examined closely enough, Richard Feynman was a multi-layered, richly intricate, complexly crafted character. No sooner had I written one thing about the man than I began to think "that's not the whole story, though . . . it just scratches the surface" or "the seeming *opposite* was also true about him." I don't know if this is more a function of the profoundly multivalent nature of the archetypes or my own Saturn-Neptune opposition, but whatever the case, this analysis proved to be an infinitely fascinating look down the rabbit-hole and a breathless, close, immediate engagement with the archetypal energies that animate the universe. Gazing at a natal chart feels akin to looking at a treasure map, every symbol a tantalizing hint at something much larger; each planet or glyph a gem of potential energy waiting to spring off the page and out of two dimensional reality into full-fledged, living, three dimensional color.

Feynman was a Sun-Uranus culture hero, a man internationally celebrated and awarded one of the highest honors our society bestows: the Nobel Prize. He was an inordinately popular and beloved teacher, a trickster mischief maker, a freedom-loving iconoclast, a ferocious intellect, and a deeply principled individual. Discovering in his natal chart the configurations of archetypal energies that gave rise to these attributes, and many that I *suspected would be there*, was an exhilarating process! Scanning back and forth between stories and footage of his life to his birth chart, it was as if the pieces of a puzzle were being put together or a fuzzy, pixelated image was coming into sharp focus. Archetypes that had been merely descriptors, albeit vivid ones, took on almost palpable form and seemed to permeate the room as I sat, hour after hour, in thoughtful reflection engaged with this man's chart. I have gained an immediate, profound appreciation for archetypal astrology, and see it as an invaluable lens through which to view the human phenomenon.

Notes

1. From the 1993 BBC Horizon documentary series "No Ordinary Genius" on Richard Feynman.

2. See Richard Tarnas's discussion of the Saturn-Neptune complex with regards to individuals who tend towards atheism and extreme skepticism. RichardTarnas, "The Ideal and the Real," 140, in *The Birth of a New Discipline—Archai: The Journal of Archetypal Cosmology*, Issue 1 (summer 2009), edited by Keiron Le Grice and Rod O'Neal (San Francisco: Archai Press, 2011). 137–158.

3. Richard P. Feynman, *What Do You Care What Other People Think?* (New York, NY: Norton & Company, Inc, 1988), 27.

4. Feynman, *What Do You Care*, 26.

5. Feynman, *What Do You Care*, 38.

6. Richard Tarnas, *Cosmos and Psyche: Intimations of a New World View* (New York: Viking, 2006), 90.

7. Jeffrey Robbins, ed., *The Pleasure of Finding Things Out: The Best Short Works of Richard P. Feynman* (Cambridge, MA: Perseus Publishing, 1999), 25.

8. Feynman, *What Do You Care*, 245.

9. Richard P. Feynman, *Surely You're Joking, Mr. Feynman!* (New York, NY: W. W. Norton & Company, Inc, 1985), 218.

10. For an excellent discussion of the dynamics behind this Saturn-Neptune impulse to poke holes in the façade of conventional reality, see Richard Tarnas's article "The Ideal and the Real" in *Birth of a New Discipline*. Here, I am attempting to describe the dialectic from a slightly different angle, namely, how it might subjectively feel for the person who has this planetary alignment in his or her natal chart.

11. See Tarnas's article "The Ideal and the Real" for an extensive analysis of the Saturn-Neptune complex, and the frequent tendency of individuals with this configuration in their charts to expose deception where they perceive it to be.

12. Wikipedia, "Richard Feynman." http://en.wikipedia.org/wiki/Richard_Feynman (December 17, 2010).

13. Feynman, *Surely You're Joking*, 21.

14. James Gleick, *Genius: The Life and Science of Richard Feynman*, (New York, NY: Vintage Books, 1992), 9.

15. Tarnas, *Cosmos and Psyche*, 93.

16. Feynman, *Surely You're Joking*, 133.

17. Feynman, *Surely You're Joking*, 62.

18. Feynman, *Surely You're Joking*, 62.

19. From the 1993 BBC Horizon documentary series "No Ordinary Genius" on Richard Feynman.

20. Gleick, *Life and Science of Richard Feynman*, 9.

21. Feynman, *What Do You Care*, 204.

22. Feynman, *What Do You Care*, 210.

23. Feynman, *What Do You Care*, 210.

24. Feynman, *Surely You're Joking*, 261.

25. Feynman, *Surely You're Joking*, 274.

26. For an in-depth examination of the Venus and Pluto archetypal principles and the multitude of ways in which they influence each other, see Keiron Le Grice's article "A Last Chance Power Drive," in *Birth of a New Discipline* on the artist Bruce Springsteen, who had the two planets in aspect to each other in his birth chart.

27. Feynman, *Surely You're Joking,* 223.

28. Feynman, *Surely You're Joking,* 188.

29. Feynman, *Surely You're Joking,* 188.

Bibliography

Feynman, Richard P. *Surely You're Joking Mr. Feynman!* New York, NY: W. W. Norton & Company, Inc, 1985.

_____. *What Do You Care What Other People Think?* New York, NY: Norton & Company, Inc, 1988.

Gleick, James. *Genius: The Life and Science of Richard Feynman.* New York, NY: Vintage Books, 1992.

"Horizon—5/7 No Ordinary Genius—Richard Feynman (1993)" YouTube video, 13:04, from BBC series documentary "No Ordinary Genius (1993)," posted by "The Chipsnbeer66," http://www.youtube.com/watch?v=NlJ19pXgHno November 16, 2010 (accessed August 11th 2011).

Le Grice, Keiron. "A Last Chance Power Drive." In *The Birth of a New Discipline—Archai: The Journal of Archetypal Cosmology,* Issue 1 (summer 2009), edited by Keiron Le Grice and Rod O'Neal. San Francisco: Archai Press, 2011. 112–136,

Robbins, Jeffrey, ed. *The Pleasure of Finding Things Out: The Best Short Works of Richard P. Feynman.* Cambridge, MA: Perseus Publishing, 1999.

Tarnas, Richard. *Cosmos and Psyche: Intimations of A New World View.* New York: Viking, 2006.

_____. "The Ideal and the Real." In *The Birth of a New Discipline—Archai: The Journal of Archetypal Cosmology,* Issue 1 (summer 2009), edited by Keiron Le Grice and Rod O'Neal. San Francisco: Archai Press, 2011. 137-158.

Wikipedia. "Richard Feynman" entry. http://en.wikipedia.org/wiki/
Richard_Feynman (December 17, 2010).

Contributing Authors

Joseph Kearns is a writer and aspiring filmmaker based in San Francisco. His work focuses on the psychological implications of revolutions in worldviews, with a special interest in developing these ideas through storytelling and the visual arts. He holds degrees from the University of Durham, U.K. (B.A., Classics) and the California Institute of Integral Studies (M.A., Philosophy and Religion).

Sean Kelly is a professor in the Philosophy, Cosmology, and Consciousness program at the California Institute of Integral Studies in San Francisco. He has published articles on Jung, Hegel, transpersonal theory, and the new science and is the author of *Coming Home: The Birth and Transformation of the Planetary Era* and *Individuation and the Absolute: Hegel, Jung, and the Path toward Wholeness*. Sean is also coeditor, with Donald Rothberg, of *Ken Wilber in Dialogue: Conversations with Leading Transpersonal Thinkers* and co-translator, with Roger Lapointe, of French thinker Edgar Morin's book *Homeland Earth: A Manifesto for the New Millennium*. Along with his academic work, Sean has trained intensively in the Chinese internal arts (taiji, bagua, xingyi, and yiquan) and has been teaching taiji since 1990. His current research interests focus on the intersection of consciousness and ecology in the Planetary Era.

Keiron Le Grice is founding coeditor of the *Archai* journal and adjunct professor in the Philosophy, Cosmology, and Consciousness program at the California Institute of Integral Studies (CIIS), San Francisco. He is the author of *The Archetypal Cosmos: Rediscovering the Gods in Myth, Science and Astrology* (Edinburgh: Floris Books, 2010), a theoretical synthesis of Jungian depth psychology, archetypal astrology, myth, and the new paradigm sciences; and the forthcoming *Discovering Eris: The Symbolism and Significance of a New Planetary Archetype* (Edinburgh: Floris Books, 2012).

He holds degrees from the University of Leeds, England (B.A. honors, Philosophy and Psychology, 1994) and CIIS (M.A., Philosophy and Religion, 2005; Ph.D., Philosophy and Religion, 2009).

Clara Lindstrom holds psychology degrees from the University of Washington (BA with College Honors, 1995, Summa Cum Laude) and the California Institute of Integral Studies (MA, 2006, East-West Psychology), where she developed an interest in the archetypal astrological perspective. She is currently a doctoral student at CIIS in East-West Psychology where her focus is embodied spirituality, the intersection of spirit and matter, and the relationship between shifts in consciousness and physiological changes in the body. Outside of academic study, Clara works with presenters to create content for CIIS Public Programs and Performances' workshops and lecture series. She is involved in ecological activism, and has a deep passion for storytelling and community building.

Grant Maxwell, assistant submissions editor of *Archai*, has served as adjunct professor of English at Baruch College and Lehman College in New York. He holds an M.A. in English from the City University of New York's Graduate Center (2011) and a B.A. in English and Plan II Honors from the University of Texas at Austin (2001). He is currently writing a doctoral dissertation in English at the CUNY Graduate Center. The dissertation, entitled *How Does it Feel?: Rock and Roll in the Evolution of World Views*, reads the music and biographies of Elvis Presley, The Beatles, and Bob Dylan in light of the work of thinkers such as Hegel, William James, Henri Bergson, Carl Jung, Alfred North Whitehead, and Richard Tarnas.

Rod O'Neal is adjunct professor in the Philosophy, Cosmology, and Consciousness program at the California Institute of Integral Studies (CIIS), San Francisco. He is a founding coeditor of the *Archai* journal and holds degrees from Vassar College (B.A., highest honors, biochemistry), UC Berkeley (M.A. , biochemistry), and the California Institute of Integral Studies (Ph.D., Philosophy and Religion). His doctoral dissertation, "Seasons of Grace: An Archetypal History of New England Puritanism," is a detailed case study of the archetypal correlations between the phenomena of a particular historical movement and the outer-planetary cycles, as well as a theoretical exploration of the ancient philosophical roots of astrology and the implications those roots hold for the current world view. Rod has been a professional astrologer for more than ten years, and has been involved in Western esoteric studies for nearly thirty years.

17757410R00136

Printed in Great Britain
by Amazon